THE COURAGE
OF BIRDS

PRAISE FOR
THE COURAGE OF BIRDS

"Kudos to Pete Dunne for a fresh take on the hardy birds that brave the cold. *The Courage of Birds* is packed with fascinating stories and tidbits. For best results, savor this book by a cozy fire with a hot cocoa."

— NOAH STRYCKER, associate editor, *Birding* magazine; author of *Birding Without Borders*

"Pete Dunne's keen and vivid prose, paired with David Allen Sibley's graceful illustrations, transforms our relationship with birds in winter. Every page offers a dazzling feat of survival. Deftly weaving behaviors with a lifetime of personal stories, the legendary author inspires us to give back to birds that brighten our lives. This book will become a well-worn companion with a special place on my desk overlooking feeders and native plants."

— MARINA RICHIE, author of *Halcyon Journey*, winner of the 2024 John Burroughs Medal for distinguished nature writing

"There is no better team in the world of birds than Dunne and Sibley, and *The Courage of Birds* showcases both their strengths— Pete Dunne's witty, insightful writing and David Allen Sibley's gorgeously evocative art—which together bring alive the winter world of birds. It's wonder and wisdom in equal parts."

— SCOTT WEIDENSAUL, author of *A World on the Wing*

"Birds in winter don't merely survive, they flourish! In *The Courage of Birds*, Pete Dunne challenges us to reimagine the colder months as a season of delight and discovery for anybody who loves to watch and wonder about birds."

— TED FLOYD, editor, American Birding Association's
Birding magazine

"Pete Dunne writes about birds as deeply respected colleagues, fellow travelers. Famously enamored of raptors, he neglects not the quail, the robin, nor the chickadee, his sense of wonder at their adaptations and capabilities fully intact. *The Courage of Birds* is biologically informed and rich with anecdote, like a good day in the field with this dean of American birders. David Allen Sibley's lovely reductionist paintings are all about form and light, unburdened by detail. It's a fine pairing."

— JULIE ZICKEFOOSE, author and illustrator of *Saving Jemima, Baby Birds, The Bluebird Effect,* and *Letters from Eden*

THE COURAGE OF BIRDS

*And the Often Surprising Ways
They Survive Winter*

PETE DUNNE

WITH ILLUSTRATIONS BY
DAVID ALLEN SIBLEY

CHELSEA GREEN PUBLISHING
White River Junction, Vermont
London, UK

Developmental Editor: Matthew Derr
Project Manager: Natalie Wallace
Copy Editor: Angela Boyle
Proofreader: Ashley Davila
Designer: Melissa Jacobson
Page Layout: Jenna Richardson

Printed in the United States of America.
First printing October 2024.
10 9 8 7 6 5 4 3 2 1   24 25 26 27 28

Our Commitment to Green Publishing

Chelsea Green sees publishing as a tool for cultural change and ecological stewardship. We strive to align our book manufacturing practices with our editorial mission and to reduce the impact of our business enterprise in the environment. We print our books on chlorine-free recycled paper, using vegetable-based inks whenever possible. This book may cost slightly more because it was printed on paper that contains recycled fiber, and we hope you'll agree that it's worth it. *The Courage of Birds* was printed on paper supplied by Sheridan that is made of recycled materials and other controlled sources.

Library of Congress Cataloging-in-Publication Data
Names: Dunne, Pete, 1951– author. | Sibley, David, 1961– illustrator.
Title: The courage of birds : and the often suprising ways they survive winter / Pete Dunne; with illustrations by David Allen Sibley.
Description: White River Junction, Vermont : Chelsea Green Publishing, 2024. | Includes bibliographical references.
Identifiers: LCCN 2024031710 (print) | LCCN 2024031711 (ebook) | ISBN 9781645022572 (hardcover) | ISBN 9781645022589 (ebook) | ISBN 9781645022596 (audiobook)
Subjects: LCSH: Birds—Wintering—Popular works. | Birds—Seasonal Variations—Popular works. | Birds—Behavior—Popular works. | Birds—Adaptation—Popular works. | Birds—Effect of cold on—Popular works.
Classification: LCC QL698.3 .D86 2024 (print) | LCC QL698.3 (ebook) | DDC 598.15—dc23/eng/20240830
LC record available at https://lccn.loc.gov/2024031710
LC ebook record available at https://lccn.loc.gov/2024031711

Chelsea Green Publishing
White River Junction, Vermont USA
London, UK
www.chelseagreen.com

To Beth Van Vleck, friend and muse

CONTENTS

PREFACE

I t was close to dinnertime. I'd spent the day hunting in the woods behind my parents' North Jersey home. The snow that had started around midday was falling heavily now, in small, fluffy flakes that veiled the comforting glow of the lights cast from our suburban home, now a mere hundred yards away. Pausing by the buffering shelter of a skirt of pin oak branches, I was startled by the sudden appearance of a Blue Jay that materialized out of the storm and fused itself to the tree's curtain of branches. The bird was close enough that I might have touched it with the barrels of the shotgun cradled in my arms. Wild birds do not typically offer this degree of intimacy, so I was brought to wonder what the jay was up to. Was it sick? Disoriented by the storm? As snow coated us, the bird studied me with penetrating eyes.

Thirty seconds later, its decision made, the jay abruptly turned, tucked its head beneath the branches, and the rest of the bird followed.

It's going to roost, I realized, and with this insight, a shiver ran the length of my body that had nothing to do with the snow or the cold. The bird had just entrusted me with its most closely guarded secret, the location of the place it would spend the night. Never before nor since have I been granted such confidence from a wild creature. But this encounter more than fifty years ago was the catalyst that gave rise to my fascination with birds in winter, and this book.

It must take great courage to be a bird, or great faith. In these pages, I extol and expound upon those virtues. In *The Courage of*

Birds, David Sibley and I hope to impart an awareness and appreciation for the millions of birds that enliven our winter landscape and instill an understanding of their needs and our obligation to them as stewards of the Earth. The primary focus of this book is the winter lives of North America's breeding birds (defined here as birds nesting in Canada, Alaska, and the contiguous 48 states north of the Mexican border—a total of approximately 695 species). It does not include the birds of Hawaii or Southern Hemisphere oceanic birds that spend the austral winter (our summer) in our continental waters. The book's special focus is the 47 hardiest of species that remain in our Arctic or Boreal region all winter (designated in this book as Tier 1 species). Also singled out for discussion are the 510 species that breed in North America and elect to remain here through the winter (Tier 2 species). Most of these short-distance migrants travel but a few hundred miles to more temperate parts of North America, including southern Canada, much of the lower 48 states, northern Mexico, and the Bahamas. The range of these hardy winter residents extends from your backyard to the rim of the Arctic ice and beyond. Far from frail creatures at the mercy of their environment, birds have proven themselves to be hardy survivors whose fortitude and ingenuity permit them to surmount winter's challenges. This book is David Sibley and my tribute to them and the culmination of two lifetimes of study that began in our youths in the natural areas close to our North Jersey and Connecticut homes. While both of us have traveled widely in search of birds, North America's winter birds continue to captivate us.

INTRODUCTION

In the summer of 2005, Warner Brothers released a French filmmaker's feature-length nature documentary that drew thousands of viewers to theaters and grossed $127,000,000 in revenue. The film was called *March of the Penguins*, and it recounted the remarkable story of the breeding strategy of the Emperor Penguin—a sturdy and stately (four-foot tall) bird specialized to face down the challenges of the long Antarctic winter to raise their young—a story of survival-against-odds, parental devotion, fortitude, and the sheer tenacity of life (all the stuff great stories, and feature films, are made of). The success of this mere nature documentary, while shocking to many reviewers, came as no surprise to those of us in the business of bringing people and nature together. While not surprised by the acclaim accorded the film, as an environmental educator and lifelong student of birds, I was somewhat miffed.

Why? I cried in my soul. Why did these filmmakers feel compelled to journey to the opposite end of the planet to document the winter coping strategies of birds when such epic dramas are played out, annually, here in the Northern Hemisphere, some as close as our own backyards?

In our hemisphere, too, birds are specialized to face down the hardships of winter. These include the Spectacled Eider, a goggle-eyed sea duck that spends the Arctic winter massed in leads of open water in the frozen Bering Sea. Another winter champion is the Rock Ptarmigan, a snow-colored grouse that endures two months of total darkness and temperatures more than fifty degrees

1

below zero. Then, there is Gyrfalcon, a robust raptor that spends the prolonged northern winter patrolling the rim of the Arctic ice cap in search of prey. And not to be forgotten, the tiny chickadees coming to your feeder engage in a daily dance with death in the hope of seeing another spring, while their dance partner, the bird-eating Sharp-shinned Hawk, struggles to do the same.

It was precisely the avoidance of predators that prompted Emperor Penguins to embrace the extreme breeding strategy that brought them recognition and fame. Yet the winsome and widespread chickadees are willing to face the challenges imposed by both predators and winter's hardships head on, and they are not alone. Nearly 700 of North America's bird species have strategies and specializations that permit them to vault seasons and move their genetic dowry forward. This book is their story: a real-life documentary played out every day as winter closes its grip over the Northern Hemisphere in your own backyard.

Welcome to winter, nature's proving ground, where there is no prize for second place during these four months. Across the Northern Hemisphere, it's "survivor take all."

You have front row seating to the greatest drama on Earth, and *The Courage of Birds* is your playbill.

BIRDS IN WINTER

Winter Wings

The origin of birds dates back 150 million years to the late Jurassic period, when a clade of small meat-eating dinosaurs (or *theropods*) took an evolutionary tack that, over the course of tens of millions of years, led to fully feathered birds with no claws on their wings but instead a single toothless bill that facilitated grasping (as a surrogate claw). Originally evolved to provide insulation, the filament-like feathers encasing early birds slowly evolved into elaborate, multipart structures that covered their bodies and, over time, enabled these next-generation dinosaurs to fly—a distinct survival advantage when 84 million years later (that is, 66 million years ago) a 10-mile-wide comet slammed into the Yucatan Peninsula, blowing a 110-mile-wide hole in planet Earth and sending a smothering cloud of soot and debris into the atmosphere that blotted out all sunlight and plunged the Earth into an impact winter lasting months, years, and (perhaps) decades. This smothering blanket sent temperatures plummeting and curtailed photosynthesis, the foundation of life on Earth. The resulting disruption to our planet's food chain caused the extinction of 75 to 80 percent of the life-forms on Earth. But birds survived.

Enabled by their smaller size, more varied diet, and (especially) the mobility of flight, birds were able to survive the mass K–T extinction while all their larger, mobility-impaired reptile relatives perished. Following the cataclysm, clearing skies and the return of

3

life-giving sunlight saw birds capitalize upon their powers of flight and embark upon a planetary-range expansion that prompted birds first to spread from South America (where birds evolved) to North America, then to Europe, Asia, Africa, and Australia via Antarctica.

But the expansion of birds into temperate regions presented these planetary pioneers with another challenge. In the tropics, life-supporting (greenhouse) conditions persist year-round, assuring birds living there of an ample and stable food supply. Above and below the equatorial zone, the Earth undergoes a dramatic seasonal shift from warm to cold, as our obliquely angled planet revolves around its sun causing sunlight to fall disproportionally across the globe, with the hemisphere that is inclining toward the sun experiencing greater hours of, and more direct, sunlight for half the year. In the Northern Hemisphere, this warming period falls between June and August (the months we call *summer*).

But when the Earth begins its Southern Hemisphere–favoring cycle (on or about June 21), colder temperatures and more hours of darkness slowly envelop the Northern Hemisphere until the period covering December, January, and February (the season we call *winter*). The onset of winter effectively ends the photosynthetic period for most plants and imposes greater stress upon birds and other animals that must, now, cope with colder temperatures, diminished food resources, and shortened hours of daylight for foraging.

Birds address winter's challenges in one of two ways. Approximately 70 percent of North America's birds fall back upon the survival mechanism that allowed them to survive the K–T disaster: they relocate to regions that are more life supporting—that is, they migrate. Some birds migrate only down to lower elevations; others like juncos and most waterfowl are short- to medium-distance migrants, relocating to more temperate parts of North America, a matter of several hundred miles travel. Still others like many warblers and thrushes retreat into the tropical zone where birds evolved. There, they team up with resident species, like

greenlets and tanagers, to form mixed-species (or foraging) flocks. And a few superachievers, the long-distance migrants, like assorted shorebird species, vault hemispheres, migrating thousands of miles, making great leaps of faith that allow these long-distance voyagers to take advantage of the seasonal bounty of the waxing austral (or southern hemisphere) summer. Only a handful of northern bird species remain in colder northern regions year-round, and they have adapted physically and behaviorally to survive winter's hardest privation, meeting the season beak-on.

Winter is not only the most dramatic of seasons. Its challenges starkly demonstrate the hardiness and resourcefulness of far northern resident birds, who are pushed to their physical limits by winter's privations. Different birds meet winter's challenges in different ways, and in sum, these adaptations tell an astonishing tale of evolutionary advancement and fortitude that pushes life to the planet's habitable rim. This book showcases and celebrates the lives of birds as they struggle to see another spring, winter's finish line. From our own backyards to the rim of the Arctic ice, birds have specialized to meet the challenges of winter. The biggest challenge to finding birds in winter is not due to any shortage of birds but to the popular misconception that our birds go south in winter so are not here to find. Most birds do go south but not far (a few hundred miles or so, not the ends of the Earth), and contrary to popular belief, in winter, 80 percent of our breeding birds remain in North America (a land mass that includes northern Mexico). The irony is that finding birds in winter is mostly easier than finding them in summer. Part of this ease has to do with distribution and part to overall numbers. For example, there are statistically millions more American Robins residing in the contiguous forty-eight states in winter than in summer. It's the robin's summer association with the omnipresent suburban lawn that imparts the impression of seasonal abundance.

Robins appear, to our minds, more common in summer because that is when they are closer to human habitation. In winter, robins largely abandon suburbia and concentrate in woodlands where they find an abundance of fruits and berries. This seasonal redistribution finds robins concentrated in orchards, grape tangles, holly groves—habitats not heavily trafficked by people.

But the seasonal redistribution of birds cuts both ways, relocating many far-flung species to places where they are more easily viewed. In summer, most North American bird species are widely dispersed, defending individual breeding territories. A good many of these species (like Say's Phoebe and Hermit Thrush) are also located in remote, often northern habitats where human observers are few.

As winter closes over the north, it squeezes birds like robins (and Say's Phoebes and Hermit Thrushes and hundreds of other North American breeding birds) to the icy rim, closer to those of us living in more temperate parts of the continent. Take again the example of the American Robin; despite their lawn-haunting predilections, American Robins breed widely across Arctic regions, too. Indeed, in summer, there is likely not a sizable willow thicket between St. John's, Newfoundland, and Kotzebue, Alaska, that does not host at least one pair of breeding robins. I once found a robin nest snugged into the corner of a crumbling sod outhouse that once served the occupants of an abandoned whaling station, a mere fifty feet from the Beaufort Sea. The nest was one hundred miles from the nearest sizable (crotch-bearing) tree and hundreds of miles from the nearest suburban lawn. And while robins breeding at extreme northern latitudes may raise only a single brood per season, across much of North America these brick-breasted thrushes produce two to three broods per breeding cycle, or two to four young per nest effort, which alludes to yet another reason winter birding can be so productive: numbers. In early winter, bird populations are at their numeric peak, with adult populations inflated by the year's crop of young.

Finding birds in winter is mostly a matter of simply getting out there and knowing where to look, insights that are much the focus of this book.

In winter, birds concentrate where they find open water and a reliable food resource. Even a trickle of water is enough to sustain a wintering Hermit Thrush, Wilson's Snipe, Killdeer, or Virginia Rail. You seek birds in winter? Just add water. Moderate rapids in an otherwise frozen river draw waterfowl the way dark-blue velvet draws lint. Hydroelectric dams and natural waterfalls that keep water from freezing attract Bald Eagles and other fish-eating birds (like herons and cormorants). Despite bitter winter temperatures, the Niagara Gorge between the United States and Canada is famous for its winter concentrations of gulls and is a seasonal gathering spot for dedicated gull watchers. The mighty Mississippi River is a celebrated winter stronghold for Bald Eagles, 2,500 of which line its tree-lined banks from St. Paul, Minnesota, to St. Louis, Missouri, on the lookout for moribund fish. The wetlands of California's Central Valley, whose daytime temperatures mostly climb above freezing, are a winter stronghold for waterfowl and shorebirds, herons and egrets.

And wherever you find birds concentrated, you'll find bird-hunting raptors. In summer, when birds of prey are nesting, they maintain large territories that are often sequestered into remote places. But in winter, Bald and Golden Eagles, Prairie Falcons, Merlins, Northern Harriers, and assorted buteo species settle in areas that are more easily accessed by most of us. If raptors are what you seek, head for any of the nation's 588 National Wildlife Refuges; many of these public lands are close to human population centers, if not actually within the jurisdictional limits of metropolitan areas, including multiple refuges in the San Francisco Bay area. The 9,000-acre Jamaica Bay Wildlife Refuge is located within the city limits of New York City and is even reachable by public transportation. The 1,000-acre John Heinz National Wildlife Refuge is adjacent to Philadelphia International

Airport, located just off busy I-95. Its wetlands, woodlands, and trail system are a resource for human and bird residents alike.

While it is true that in winter most birds are not advertising their presence with song (a boon to bird finding in other seasons), birds in winter are far from mute and many offer up an array of vocalizations that serve a variety of functions. Some of these calls rank among the most evocative and iconic sounds in nature.

No walk along a winter beach would be complete without the keening cry of gulls. The brassy bray of jays is a signature sound across many a winter landscape. Perhaps no sound is as wild and haunting as the whooping sigh of the Tundra Swan, a species that, as the name suggests, breeds far from most humans but winters by the thousands in the back bays of North Carolina's Currituck Sound and the Northern Sacramento Valley of California. And almost anywhere you journey, the cheery *dee, dee, dee* call of chickadees can be counted upon to enliven a winter day.

In many places, the harmonic two-note bark of a Canada Goose is the quintessential sound of late autumn and winter. Emanating from echelons of high-flying birds, it harks to places that lie beyond the horizon but within range of human longing.

Another celebrated winter vocalist is the Great Horned Owl, that most familiar and widespread of owls. These birds are actually most vocal in winter, when they begin their breeding season. Every December, along a well-forested neck of land not far from my South Jersey home, I can hear up to five pairs of owls duetting from across the marsh, and our resident pair's favorite perch is the Methodist Church steeple, right next to our Mauricetown home. I'm not suggesting the owls are Methodists—as far as I know owls are nondenominational—but this steeple is the highest point in town and the perfect place for the owls to get their point across, which is, "Sorry, Bub and Bubbette, this territory is ours. Yours is somewhere else. Beat it."

In Great Horned–ese, the message is rendered, *Who, who, who, who. Hoohoo.* The Great Horned's territorial posting does not apply to the Eastern Screech-Owls that reside in our neighbor's tree-rich yard across the street, nor to the Barn Owls that overfly the town. Owls as mostly permanent residents across much of North America are part of winter's bird-rich dowry—special prizes for human observers hardy enough to weather winter's chill and embrace the season and its abundant bird life. This seasonal dowry includes, at times, pearls of great price, like the regal Snowy Owl, the Holy Grail of winter birds and one whose periodic "invasions" are bound to make the local news and draw a crowd of onlookers. Wintering regularly as far south as the Central United States, these large snow-colored birds with the cat-like eyes spend their days scanning for prey in mostly treeless terrain. They are easily viewed but are also easily dismissed as just another beached Clorox bottle or a dirty mound of snow, until the great birds swivel their heads and turn their piercing yellow eyes upon you. Catch. You're prey.

"Migrate or Tough It Out?" That Is the Question

Both survival strategies are perilous, and attrition is high either way with 60 to 90 percent of birds that hatch in any given year failing to see the next. In migration birds pay forward, embracing greater risk and great physical stress in the short run, in return for greater comfort and food security in the long. But then migrants must face a perilous return trip.

Birds that overwinter in the far north face a grueling day-to-day struggle for survival for months and court the perpetual risk of periodic, balance-tipping hardships (like a prolonged freeze, a crippling blizzard, or a depleted food supply).

Neither survival strategy evolved spontaneously. As an animal group that evolved in the hot, humid environment characterizing the late Jurassic Period, protobirds did not simply emerge from the egg equipped to vault hemispheres or survive subzero temperatures. These adaptations took hundreds of thousands of years to evolve and much trial and error. If you've got what it takes to survive, you get to pass your genetic dowry on—the gold ring at the end of winter's travail. Or you fail, you die, and you take your evolutionary shortcomings with you. Somebody else gets the ring. But today, there are bird species that routinely make nonstop overwater crossings of several thousand miles, and birds that are adapted to survive temperatures that would make a parrot-cicle of your average Amazonian conure in an Arctic minute.

The planet's migration champion is the Arctic Tern, a slender-winged seabird whose 45,000-mile circumnavigation of the Atlantic Ocean finds birds spending December 1 to April 1 in the food-rich waters off Antarctica, the opposing pole of their bipolar lives. The tern's passage to and from breeding and wintering areas consumes four months of the year, meaning a third of an Arctic Tern's life is spent on the wing. By this accounting, the Arctic Tern might better be known as the Pelagic Tern, a bird whose defining habitat is open ocean.

A Tale of Two Falcons

Among the Arctic's 280 or so breeding bird species are two large falcons: the Peregrine and the Gyrfalcon. Both are cliff-nesting species that may even alternate favored nest ledges one year to the next. But these two mostly bird-hunting raptors could hardly be more different in terms of their wintering strategies. Tundra Peregrines, preying mostly upon shorebirds, vacate the Arctic entirely after the breeding season, using much the same elliptical migration route as their shorebird prey. This perilous journey takes many Tundra Peregrines out over the Atlantic Ocean and as far south as middle South America, a minimum distance of 15,500 miles. Departing in September and returning in May,

Tundra Peregrines never experience the two months of darkness and intense cold of the Arctic winter night. Their nonbreeding months are spent in mostly coastal locations, overlooking beaches and mudflats where nonbreeding shorebirds concentrate, and the falcons enjoy the warmth of a tropical sun on their backs.

Gyrfalcons are different, with those birds breeding below 70 degrees north latitude (Fairbanks, Alaska, lies at 65 degrees north latitude) and remaining on their breeding territories year-round. More northerly breeders evacuate (somewhat) with a few Gyr-falcons regularly wintering as far south as southern Canada and northern border states. But many, if not most, of these Arctic falcons appear to lead nomadic (even pelagic) lives in winter, patrolling the rim of the seasonal ice sheet in northern seas for weeks on end, hunting alcids, gulls, and hardy sea ducks; even roosting on the pack ice or on offshore ice floes. The birds may spend up to forty days away from land and then return to nest ledges while snow still covers the landscape, switching their diet back from seabirds to the Gyrfalcon's summer prey of choice, ptarmigan. Arriving weeks ahead of their peregrine neighbors, Gyrfalcons get first choice of prime nest ledges—typically favoring those cliffs that offer a warming southern exposure and ideally a protective overhang.

As a winter nomad, Gyrfalcons may navigate winter "territo-ries" up to 40,000 square miles and log nearly 3,000 air miles in search of prey, much of it over open or frozen seas in near total darkness and subzero temperatures—quite a feat for a member of a bird family most closely related to the parrots.

The Gyrfalcon's larger size (body mass) helps it withstand the bitter cold of the Arctic winter (Bergmann's rule). And while some populations of Alaskan peregrines are also nonmigratory, most notably the Peale's subspecies (*Falco peregrinus pealei*), which breeds and winters along Alaska's southeast coast and offshore islands. The Arctic-breeding Tundra Peregrine (*Falco peregrinus tundrius*) is wholly migratory. As might be expected and in accordance with Bergmann's rule, the Peale's Peregrines are also somewhat larger than Tundra peregrines; in fact the larger female Peale's Peregrines are close to male Gyrfalcon size. As an added protective safeguard, Peale's Peregrines have a dusting of water-shedding powder on their outer feathers—a waterproofing refinement that helps keep the birds dry in the supersaturated environment they call home. Dutch Harbor, Alaska, on the Aleutian Islands, receives sixty inches of precipitation per year and ninety inches of snow, making it one of the wettest places on Earth but a place that Peale's Peregrines are supremely suited to exploit—a region rich in cliff-nesting alcids and other seabirds.

Built for the Cold

Unlike their reptile ancestors, birds are warm-blooded, so able to withstand temperatures that immobilize cold-blooded animals. Like mammals, and unlike reptiles, birds also have efficient four-chambered hearts. Bolstered by their high metabolism and advanced circulatory systems, birds are thus able to maintain body temperatures of 102 to 109°F. Birds can also generate emergency heat by shivering on command. Some, like waterfowl, have ingenious circulatory modifications in their unfeathered feet and legs to reduce heat loss.

But for the most part, birds maintain their body temperatures by feeding voraciously as temperatures fall. Food is fuel to fire the bird's metabolic engine. In extreme northern latitudes, it is restricted foraging time, more than food availability, that determines the northern limit of a species' winter range. Above the Arctic circle, in December and January, daylight is restricted to a mere two hours of twilight per day and outside temperatures habitually dip to −50°F. Very few birds have adapted to meet these challenges, but it is these Arctic super achievers that have pushed life to the very rim of existence and constitute one of the most amazing chapters in the story of life on Earth. Dinosaurs, it turns

out, did not become extinct; they merely evolved into birds to better accommodate changing and more challenging conditions. They did so mostly by evolving feathers, the defining characteristic of birds and their first line of defense against winter cold.

Feathers: The Evolutionary Edge

Feathers both enable the great migratory flights of birds and protect birds from the elements. While structurally sophisticated in design, feathers still wear and must be replaced. A number of long-distance migrants replace flight feathers prior to or after migration in order to fly at peak efficiency.

In addition, many bird species have a colorful breeding plumage and a more cryptic nonbreeding plumage. It is a bird's flight feathers (the longer, stiffer wing and tail feathers) that permit birds to vault hemispheres and move from food-impoverished areas to places offering greater food security.

The engineering marvel that is a feather is not only unique to birds but defining insofar as birds are the only animal group clad in feathers. Feathers' many variations allow birds to blend in, stand out, stand in the rain and remain dry, stay warm in temperatures that would daunt most mammals. And they permit birds to fly—an engineering feat our own species achieved only a little over a century ago.

Assemble the planet's finest structural engineers and challenge them to design a strong, durable, all-purpose covering that will protect, warm, support flight, and be as "light as a feather." They could hardly come up with a design superior to the feather.

Constructed of carotin, the same substance as your fingernail, these highly evolved dinosaur scales also appear to have covered some dinosaur species (so called "dino fuzz"). Each individual feather consists of a central cylindrical shaft (or rachis). Radiating from the shaft are two paired and opposing rows of barbs, each of which supports a finer array of barbules, which are designed

15

to interlock tightly with adjacent barbules via Velcro-like hooks, creating a near-seamless outer surface. When fresh, groomed, and lubricated with the thick oily substance secreted from a bird's uropygial (oil) glands, feathers make water bead up like water (rolling) off a duck's back.

Feathers are strong yet elastic, and replaced more or less regularly. Laid down in tracts, each feather is close enough to the next to overlap with their neighbor and tracts are close enough to present a weatherproof, shingle-like barrier to the world. Between the tracks are fluffy filament-rich "down feathers" that trap warm air close to the bird's skin. Air, with its widely spaced molecules, is an excellent insulator.

It is these underlying, heat-trapping down feathers that permit resident species to remain in the far north and that give birds the ability to defy temperatures beyond the capacity of other animals, including most fur-bearing mammals. Encased in their waterproof down jackets, birds can be nearly impervious to winter.

Eider down, with its dense array of heat-trapping filaments, is the warmest natural substance known to humankind. A pinch of eider down placed on a forearm is registered as heat not weight. Encased in their eider-down armor, birds like Spectacled Eiders can thrive in temperatures of −50°F. The outer body-feather layer and folded wings help keep the underlying down feathers dry and add an extra layer of warmth.

Feathers, as inert appendages, cannot repair themselves. They wear and abrade. Birds solve this problem by molting worn feathers and replacing them with new ones. Most birds molt once or twice a year with mid-summer and early mid-winter constituting prime replacement periods. By molting in mid-summer and mid-winter, migrating birds face the rigors of spring and autumn migrations with fresh feathers operating at peak efficiency. Some species (like most waterfowl) choose to molt all their feathers at once. Other species that need to fly to forage (like birds of prey) molt their feathers gradually and sequentially, so while they might not be operating at peak efficiency, these gradual molters never entirely lose their power of flight. Birds that winter in northern regions also typically produce more down feathers during their mid-summer molt in preparation for colder days ahead.

The longer wing feathers that facilitate flight are contoured to form a curved upper-wing surface, a vacuum generating air-foil that, when sliced through the air, produces lift. Overlying the base of these primary and secondary feathers are feathers called coverts that create a tight, lump free surface so air can flow smoothly across the wing. To maintain peak efficiency, feathers demand daily maintenance, an exercise called *preening*. Periodic bathing (in water or snow) even in winter keeps feathers operating at peak efficiency. Preening removes dust, dirt, and parasites, and birds may spend most of their daily time budget on this critical exercise.

Other Cold-Defeating Mechanisms

Fluffing: When temperatures fall, birds can increase their insulating layer's depth (or loft) by fluffing out their feathers, turning themselves into feathered puffballs.

Feeding: Ten percent of a healthy songbird's body weight is fat. In winter, birds spend much of their daylight hours foraging for fat-rich foods that contain twice the heat-generating energy of proteins or carbohydrates.

Head-tucking: By tucking their heads and bills beneath their scapular feathers, birds reduce heat loss and recycle warmth in the same way people do when breathing into cupped hands. Many birds also perch on one leg, tucking the unused, unfeathered leg into their body feathers and further reducing heat loss.

Sunning: Turning your back to the heating rays of the sun is an excellent way to boost body temperatures. Of course, this technique demands sunlight, which in mid-December in the Arctic is in very short supply.

Cuddling: Many bird species cluster or perch side-by-side, or gather in tight flocks, sharing body warmth.

Flocking: Among flocking species like shorebirds and gulls, the outermost birds block heat leaching winds, offering interior flock-mates greater protection. This is the same heat conserving technique used by breeding Emperor Penguins in Antarctica.

Roosting: At night, birds seek out protective confines, like dense vegetation or tree cavities, adding another outer barrier against the cold. Going to roost early helps birds get prime perches, reduces heat loss, and conserves energy.

Torpor: Some bird species, like chickadees and hummingbirds, are able to induce a state of reduced metabolism, saving energy during the night by reducing their metabolic needs.

Shivering: Birds can generate emergency heat by shivering on demand, a mechanism that produces heat through muscle action, much as we do.

Burrowing: In snow-covered landscapes some roosting birds, like ptarmigans and Gray-headed Chickadees, may burrow beneath fluffy heat-trapping snow, which is mostly air.

Camouflage: Becoming One with Winter

The ultimate winter survivalists are arguably the Willow and Rock Ptarmigans, grouse species specialized to endure the long Arctic winters. Mostly brown in summer, these compact tundra resident species molt into wholly white body plumage in the fall, rendering them nearly invisible against the winter snow. In addition, in winter, grouse toes sprout stiff pectinations, turning their feet into tiny snowshoes. But replacing feathers is energetically expensive, siphoning off nutrients needed to meet a bird's metabolic needs.

Snow Buntings, whose plumage is brownish in summer and fall and mostly white in winter, have devised a more energy-efficient way of changing into dress whites. While somewhat migratory, our northernmost breeding passerine returns to the

Arctic while snow still blankets the ground. They scrub their bodies against the snow to broom off brown-tipped body feathers until they abrade down to their snow-white base. Tucking their grass, moss, fur, and feather-lined nests into protective rocky crevasses, the birds keep their two to seven eggs viable in temperatures that may dip to −22°F. Typically, Snow Buntings have time for but a single brood per nest cycle, so nest failure is not an option. Both adults attend the young until they fledge, feeding them mostly the insects that flourish in the Arctic summer.

But Why of All Places the Arctic?

People who have never visited the Arctic in summer have the image of a barren, lifeless place. Nothing could be farther from the truth. During the warmer months, May to August, the tundra comes alive—a huge life-filled terrarium brimming with flowering plants, emerging grasses, and sedges, and flush with bumblebees, butterflies, moths, and mosquitoes in such numbers that caribou are sometimes driven to run to escape the tormenting cloud of blood-sucking insects. With temperatures sometimes climbing into the 70s (averaging in the 50s) and twenty-four hours of sunlight, the summer tundra is one of the most life-rich places on Earth. Young birds, needing protein, grow quickly in this insect-rich environment and fledge in time to beat the clock ticking down to winter. Shorebird chicks are so precocial and food so abundant that most young require no more from their parents than protection from the elements and from predators. Most shorebird chicks are able to walk and feed within hours of emerging from their eggs.

Indeed, among shorebirds it is common for one parent to depart as soon as chicks hatch, leaving all parental care to the remaining adult. This care-giving adult, too, typically migrates before young fledge, flying on to food-rich staging areas where they fuel up for the migration ahead by larding on layers of energy-rich fat. Young birds follow adults soon after fledging, but this staggered migration timetable (adult, other adult, then young) reduces competition for food at prime staging areas by allowing sex- and age-segregated waves of birds to use the same food-rich feeding locations at different times. By September, almost all shorebird species have left the Arctic. At these high latitudes, snows can begin falling in August, so dawdling is not an option.

Migration: The Better Part of Valor

Avoiding winter is probably, in the long run, the best survival strategy for northern birds when facing the approach of winter. And most of North America's 700 or so breeding bird species do take to their wings and relocate, at the very least, to places offering greater food security and diminished hardship. These migrant species range from the American Dipper, an aquatic songbird breeding north to Alaska's Brooks Range, to the Northern Wheatear. Dippers are permanent residents of the cold, clear mountain streams, but they may migrate to lower elevations as winter closes in, still remaining within the same watershed. At the opposing extreme of the migratory scale is the Northern Wheatear, a small, thrush-like bird that breeds on rocky tundra slopes in Alaska and northern Canada east to Greenland, and winters in Africa using either a trans-Atlantic or pan-Asian route. Those Alaskan wheatears taking the pan-Asian route must navigate a 9,000-mile journey that requires three months and forces birds to cross the resource-deprived Arabian desert. (To appreciate this challenge, think back to the movie *Lawrence of Arabia* and the attack on Aqaba.) After a few short weeks in East Africa sharing the rocky plains with zebras and wildebeest, wheatears then retrace their routes, arriving back in Alaska while musk ox are shedding their insulating layer of qiviut, this Ice Age mammal's super-warm, inner wool layer.

By far, most North American breeders migrate just a few hundred miles, staying for the most part within North America.

These include many of the species visiting your backyard bird feeders, like juncos and other assorted sparrows.

But those species that do migrate great distances must prep for their journey, replacing flight feathers, laying down layers of fat to burn for fuel, and even, in some cases, shrinking the size of some internal organs in order to reduce weight. These long-distance champions include the Swainson's Hawk, Blackpoll Warbler, and Sanderling.

A large, slender-winged prairie buteo, Swainson's Hawk is a widespread breeder across the western prairies north to Alaska that winters in the grasslands of Argentina, Paraguay, and southern Brazil. Flying a minimum of 6,000 miles (one way), these mostly insect-eating grassland raptors are obligated to fly over the Amazon Basin, whose densely forested habitat is unsuited for a bird of prey specialized to capture large insects on foot.

Fattening up before their long journey by gorging themselves on grasshoppers, and then relying mostly upon energy-conserving thermals for lift, hundreds of thousands of Swainson's Hawks make the two-month journey from the United States to the Pampas, apparently fasting the entire way.

An obligate flocking species, Swainson's Hawks are routinely recorded at migratory choke points like Corpus Christi, Texas; Vera Cruz, Mexico; and Panama, from August through October. Flying thermal to thermal, the birds coalesce into large swirling "kettles" (flocks that may number hundreds of birds) and ride the bubbles of air aloft, then, using gravity for propulsion, they trade altitude for distance by setting their wings and gliding to the next thermal. Leaving in September while the sun still sits high in the sky, the birds are assured of finding strong and abundant thermals along their entire northern route. Flying in massed numbers enables birds to easily target thermals lying ahead by noting concentrations of birds massed along the flight path.

The Blackpoll Warbler is a small (46-ounce) boreal-forest breeder that makes one of the planet's most daring migratory leaps. Jumping off northeastern Canada, the fat-laden birds make an 11 to 12,000-mile, 4-day nonstop flight out over the Atlantic to wintering grounds in northern South America, south to Peru and Brazil. To reach their North American point of embarkation, some birds must first fly from Alaska clear across Canada.

Conducted between August and October, the birds are obligated to migrate over open ocean through the Atlantic hurricane season. Flying day and night, at altitudes up to 5,000 feet, for up to 72 hours, the tiny birds arrive in South America famished, their fat reserves depleted. Some birds on their final approach are reduced to metabolizing muscle tissue—the very engines that bear them aloft.

––––

The Sanderling is known to most of us as the tiny sandpiper that plays tag with waves on the beach. Few beach strollers realize that the hyperactive birds do not nest locally. In June, Sanderling are mostly absent from our beaches because they are, then, nesting in the High Arctic. Failed breeders, still sporting touches of their rust-colored breeding plumage, begin their return to southern shores in July. Successful breeding adults follow in August, and spangle-backed juvenile birds follow in August and September. Wintering on sandy beaches from southern Alaska and southern Maine, south to Chile and Brazil, Sanderling have one of the longest latitudinal winter ranges on Earth. Other hemisphere-jumping shorebird species include the closely related Red Knot, which winter as far south as Tierra del Fuego, but Sanderlings hedge their bets, with some birds wintering in the Northern Hemisphere and some in the Southern. Why some individuals remain north and others are willing to accept the risk of a longer migration is a mystery, but it may simply be nature's way of spreading out the risk, a species hedging its bet against the perturbations of a hostile-to-indifferent universe by putting their chips on both red (Northern Hemisphere) and black (Southern Hemisphere).

Another shorebird with an exceedingly long and narrow winter range is the sturdy Surfbird, a plump shorebird species that breeds on rocky Alaskan alpine tundra, and winters along rocky seacoasts from southern Alaska to southern Chile, where it forages within the "splash zone."

Not all birds use their wings to migrate or, more accurately, to fly. The Atlantic Puffin, a chunky alcid, breeds on rocky north Atlantic coasts from Maine to Greenland, and winters at sea over continental shelf waters. In August, after the breeding season, adults and young swim seaward, using their wings for propulsion and their feet as rudders, much as penguins do. Traveling as far south as Maryland, the bulbous-billed seabirds seek out concentrations of small fish. Indeed, puffins spend most of their lives at sea, coming ashore only to breed. Other alcids, too, winter at sea, mostly in northern waters, but Razorbills may travel as far south as Florida.

Migration: How Far Is Far Enough?

Migratory distances differ widely from species to species: from the epic sagas sweeping hemispheres to some birds migrating almost locally, such as the alpine-breeding rosy-finches, migrating only from higher altitudes to lower, and the High Arctic–breeding Snow Bunting, migrating only as far south as coastal beaches in North America. Many waterfowl species migrate only far enough to be confident of finding open water. It is among the shorebirds that we find migrations vaulting hemispheres.

These long-distance voyagers include a migratory champion, the Bar-tailed Godwit, whose nonstop 7,000-mile transpacific migration ferries birds to and from Alaska and New Zealand or southeastern Australia twice a year. This species is the current record holder for nonstop flight when, in 2022, a radio transmitter–bearing juvenile Bar-tailed Godwit (designated B-6) flew from southeastern Alaska to Tasmania—a nonstop 8,435-mile flight. This transpacific traverse, which spanned October 13 to October 24, passed west of the Hawaiian Islands and overflew both the Marshall and Solomon Island groups. From takeoff to touchdown, the bird was in the air eleven days and one hour—a remarkable feat of endurance and evolutionary refinement.

While some birds, like the Bar-tailed Godwit, migrate in a single nonstop leap of faith, others fly in calculated stages, their arrivals at key staging areas timed to coincide with some seasonal food abundance that permits migrants to refuel quickly and get back on their way. In spring, on their return flight, Bar-tailed Godwits adopt this more protracted strategy, making an interim stop on the critically important, food-rich mudflats of China's Yellow Sea, where birds may remain for several weeks, laying down the fat reserves they will need to complete their journey back to Alaska. They must arrive there in prime physical condition with fat still to burn at a time when local food resources remain in short supply.

Still other bird groups, like gulls, are nomadic, wandering all winter in search of food. Short-distance migrants, like waterfowl, remain for the most part in North America (including Northern Mexico and the West Indies). But many warbler species relocate to Central America, where they join resident birds like greenlets and tanagers in mixed-species flocks that move through rain- and cloud forests like a hungry cloud in search of insect prey.

Many short-distance migrants don't push the envelope, traveling only as far south as they deem necessary. Sometimes they goof, wintering farther north than conditions can provide for. I live in a region where both American Woodcock and Great Blue Heron overwinter. In warm winters, they survive; in colder winters when ponds and forest floors freeze or become snow covered, they perish, taking their poor judgement to the grave with them. It's all about the food. Birds congregate where they can find dependable and accessible food reserves. This means different resources for different species. In winter, many birds of prey, like harriers and owls, congregate where rodents are experiencing a population boom. Rodent numbers are cyclic, high one year, low the next. In a year marked by plenty, hundreds of Short-eared Owls may congregate in vole-infested grasslands in Oklahoma, only to be all but absent the next when rodent numbers crash. The owls don't perish, they simply go elsewhere—on to the next rodent bloom.

Blue Jays, which are mostly resident, vacate northern woodlands by the thousands in years when the local acorn crop falters. Migrating in large low-flying flocks in daylight, the birds are on the lookout for acorn-rich forests to support them. After winter quarters are located and the birds are established, they busy themselves burying acorns for later retrieval—a strategy used by many wintering species called "caching." You've noticed this behavior at your bird feeder among chickadees that perch momentarily, then fly off with a seed in their bills.

Winter finches like crossbills, Purple Finch, Pine Siskin, and Evening Grosbeak are essentially nomadic, wandering through northern forests in search of seed-bearing trees. As winter progresses and food stocks become regionally diminished, the flocks move on. Pine Grosbeak numbers are contingent upon mountain ash berries. Poor berry year? Grosbeaks head south, in numbers staging "irruptions." In winter, Common Redpoll, an abundant Arctic and subarctic breeder, congregate where they find a surfeit of birch catkins. An ice storm across New England that encases catkins in bill-defeating ice may send flocks of redpoll farther south even as late as February, a time when many species are preparing to migrate north. Red-tailed Hawks winter, in numbers, as far north as they can find mostly snow-free ground that allows them to easily spot prey. A heavy dump of snow in February in northern New York will propel birds south long after their normal fall migratory period (August through

November) has ended. If the winter is cold enough for the Great Lakes to freeze, wintering Red-necked Grebes (a diving, fish-eating species) have no choice but to head for the unfrozen New England seacoast or to some swift-flowing, unfrozen river.

The Terrible Winter of 1977

While December and January are offseason for human visitors to the Jersey shore, winter is prime season for waterfowl. One of these seasonal regulars is Atlantic Brant, a small marine goose that resembles, somewhat, a small, compact Canada Goose. Breeding coastally in the High Arctic, wintering coastally from Cape Cod, Massachusetts, to the Outer Banks, North Carolina, the hardy birds have traditionally subsisted on eelgrass and sea lettuce that they pluck from shallow back-bay waters. The species' wintering stronghold has always been the bays of coastal New Jersey and New York, where 90 percent of wintering Atlantic Brant are found. Typically, moderate winters ensure the birds an adequate food supply and high survival rate. Salt water rarely freezes at this latitude. Then came the winter of 1977, the coldest on record in the Northeast.

The first challenge to wintering brant that year came in August when Hurricane Belle (a Category 3 storm) churned its way up the Atlantic coast, uprooting the all-important sea lettuce beds and sweeping the plants out to sea. Then in January, the Northeast began being pummeled by a relentless parade of Arctic cold fronts. The normally temperate Cape May, New Jersey, endured 86 percent of its days below freezing with the thermometer sometimes dipping below 0°F and never climbing above 51°F. First the freshwater lakes froze, then the back bays and channels. Even Delaware Bay was locked in ice, curtailing ferry service. The buried water main serving the town of Cape May Point, too, froze, leaving residents without water. I was one of these, and about two weeks into the freeze, I took a walk

down the beach between Cape May Point and the city of Cape May and stumbled upon several brant immobilized on the sand. Healthy wild birds do not allow close approach, but these cold-stressed birds had no choice, their breast feathers frozen to the sand. Extracting one bird with a minimum of feather loss, I examined its breast, which even through gloved hands I determined to be wasted down to the keel. Clearly the bird was starving. Cradling the bird in my arms, I drove it to the Wetlands Institute, a marine laboratory whose wet lab had been converted into an emergency rehabilitation station for starving brant. Dozens of birds arrived daily, ferried there by everyday citizens moved to pity by the bird's plight. The challenge was the specialized diet of brant. Sea lettuce is not an item commonly stocked on the supermarket shelf. We tried shelled corn (relished by Canada Geese), turkey pellets, even several varieties of lettuce off the grocer's shelf, but having no experience with these food items, the brant stood amid our offerings and simply starved. As far as these highly specialized feeders were concerned, we might as well have been offering them sand.

All we could provide the birds was shelter from the bitter cold. Every morning, new corpses were added to the mound of birds outside the door. Multiple government agencies and assorted not-for-profit organizations tried airdropping corn to isolated flocks with the same failed results. These specialized birds needed sea lettuce or eelgrass, which the frozen back bays had removed from the menu. Ultimately the scouring winds blew the six-inch snowpack blanketing the Cape off recreational sports fields, and in desperation, birds began nibbling the exposed dry grass. The expansive Cape May Coast Guard training-center parade ground supported hundreds of birds. The bitter cold lasted into March, and New Jersey–state waterfowl biologists flying aerial surveys estimated that the wintering brant population fell from a high of 851,000 birds to 28,000, a staggering decline that cannot begin to consider the suffering endured by the birds themselves. Even many American Black Ducks, that hardiest of dabbling ducks, starved, as did assorted other water-bird species. But as tragic as the winter of '77 had been for waterfowl, periodic natural calamity is built into the system. The occasional harsh winter is part of nature's long-term survival plan, a vetting process that supports stronger birds, improving the species' survivability in the long run. Fifty years on and, following decades of typically moderate New Jersey winters, wintering brant numbers are once again over 250,000 and the population appears mostly stable. But the bitter winter of 1977 will go down in the record books as an example of the knife-edge that wintering birds straddle.

The silver lining in the disaster were the many hundreds of individual people who were brought together by the plight of the birds. And it seems that the brants learned a new survival trick that may serve them in the future. Following that calamitous winter, a few wintering brant can now always be found foraging on grassy fields and along roadside medians, grazing on withered grass, even though sea lettuce beds have recovered and the honking *purr* of brant once again resonates in the back bays of New Jersey, as much a part of the winter soundscape as the keening cry of Herring Gulls and the gruff chortle of the predatory Great Black-backed Gulls, who fed well that terrible winter of 1977. It's an ill wind that doesn't blow some gull some good.

Migratory Irruptions

Food shortages are related mostly to the cyclic nature of food stocks or to natural calamity, such as fire, storm, or drought. In times of privation, many seed-eating species engage in irruptive migrations. These massed relocations of birds abandoning food-impoverished regions in favor of more food-rich regions are routine among several bird families, most notably the finches.

During irruption years, triggered for the most part by cyclic shortages of pinecones or acorns, even mostly sedentary species like Blue Jays and Boreal Chickadees may evacuate northern forests in greater numbers. Birds like the Pine Grosbeak (North America's largest finch) will migrate farther south and in greater numbers than usual when mountain ash have a poor fruiting year. Food shortage affects flesh-eating birds, too. American Goshawk, a large, formidable raptor of mature northern forests, famously stage widespread winter "invasions" involving hundreds to thousands of individual birds when the cyclic populations of grouse and snowshoe hare simultaneously reach their low points.

But at least one northern raptor's "invasions" are triggered not by a shortage of prey but abundance. This is the Snowy Owl. Typically, the female lays one to nine eggs per year. But in years marked by high lemming numbers, the Arctic breeders may produce a surfeit of young (up to eleven owlets per nest). It is this abundance of first-year birds, traveling south in numbers, that constitutes the bulk of these winter invasions. The most recent northeastern Snowy Owl invasion was the winter of 2013 to 2014—a movement that propelled some birds as far south as Bermuda and Florida and whose nearly unprecedented numbers even made the pages of the *New York Times*.

Snowy Invaders

When I was in 7th grade our science teacher, Mr. Trato, informed the class that there was "no such thing as cold." Cold, he explained, was simply the absence of heat. I could not accept this then; I do not accept it now.

As a lifelong student of winter, I assure readers, that the laws of physics notwithstanding, cold is a real tangible entity, a physical manifestation as real as wind and rain and ice. The first time I stepped into 30 degrees below zero air, it was like being hit in the face with a Louisville slugger. It was physical. It was real. And it hurt.

The Snowy Owl, while a spiritual representation of winter, is a real, physical manifestation, too, a flesh and blood and feather embodiment of the season—a bird tempered by the cold to master the greatest challenges winter has to offer. Breeding across Arctic regions north of the taiga forest, Snowy Owls lead (for the most part) nomadic lives. Breeding where lemmings abound, the birds may journey over a thousand miles from a previous year's nest site to settle where lemming numbers are high, ensuring a successful breeding season. The female owl's maternal instincts are spurred by the volume of lemmings the male brings to the nest. He keeps bringing in the rodents, she keeps mating. Win/win. And brood success during years of plenty is correspondingly high, with more young surviving to fledge.

After the breeding season, adult females typically remain in the Arctic. Juveniles disperse widely, many (particularly juvenile males) venturing south, some as far as the Canadian Prairies and northern States. Years marked by high owlet productivity, every three to five years, precipitate minor irruptive flights (or Snowy Owl invasions). Major irruptions, involving many hundreds of owls, occur every century or so and propel birds well south of their normal winter range where they attract widespread attention.

Long before Harry Potter's Hedwig captured reader's hearts, and before scientists understood the driving mechanism behind the Snowy Owl's periodic invasions, the great white birds were celebrated as ambassadors of winter. In Inuit culture the Snowy Owl is associated with guidance and wisdom. Thirty thousand years ago, our ancestors painted their likeness on the walls of caves in southern France, right next to depictions of now-extinct horses and lions. Today, the Snowy Owl is the official bird of Quebec, Canada, and a favorite subject of winter holiday greeting cards.

After the breeding season, Snowy Owls begin wandering south, some as early as November. During irruption years, birds may remain in the south until May, although typically birds begin their return to Arctic breeding grounds in March.

THE COURAGE OF BIRDS

Only the adult males are wholly white, females and juveniles are richly barred with dark banding and recall soot-flecked (even soot-coated) snowmen. At any time of year, this northern owl is rarely far from snow. Breeding as far north as winter snowpack permits, the Snowy Owl is admirably suited to exploit the winter environment. Overall rotund and compact, the mass-to-surface-area ratio is ideally tailored to conserve body heat. Beneath the bird's dense layer of feathers lies North America's heaviest owl, at 4.5 pounds, a pound heavier than the Great Horned Owl and twice the weight of the Great Gray Owl (another hardy, far northern resident). Unlike the legs of most birds, the legs of Snowy Owls are feathered to the toes, which the birds attend to with cat-like attentiveness. However, the bird's greatest heat conserving device may well be its sedentary nature. Accomplished predators, Snowy Owls can dedicate much of their daily time budget to placidly perching and scanning for prey. Favored perches include sand dunes, utility poles, haystacks, beach homes, and, in northern seas, icebergs.

Almost any open terrain may support Snowy Owls in winter, but their typical habitats include prairie, open farmland, and marshes, as well as upper beaches, lake shores, and coastal dunes. In winter, some of these owls haunt polynyas or unfrozen portions of large lakes where ducks and other water birds concentrate. John James Audubon once described a Snowy Owl attempting to catch fish on the edge of one such opening. More typically, in winter, Snowy Owls hunt prey ranging in size from rodents to geese with Mallard-sized ducks being especially prized. Radio tagging has demonstrated that some wintering birds lead mostly sedentary lives, wandering no more than a quarter mile from their established core winter range, while other individuals are nomadic, traveling hundreds of miles over the course of a winter, touching down in assorted locations, including coastal communities, urban centers, and major metropolitan airports where they constitute a hazard to

32

air traffic and must be physically removed. Almost everywhere Snowy Owls occur, they are bound to attract attention and become a media sensation. The unpredictable nature of their movements only enhances their mystique, and their periodic invasions are the source of legend: The massive 2013–2014 northeastern invasion made the papers from Boston to New York and engaged multiple TV news crews out to get the "owl story." This mega-irruption, the largest in over a century, involved hundreds of owls from Wisconsin to New England, south even to northern Florida. The last invasion of this magnitude occurred between 1876 and 1877, and museum specimen trays in assorted ornithology departments still bear evidence of this celebrated incursion.

Though daytime hunters, Snowy Owls become increasingly animate at dusk. When prey is sighted, the owls fly low and directly, taking advantage of the contouring terrain as well as their broad wings and silent flight to surprise prey.

During normal winters, a few birds can be always counted upon to occur as far south as Long Island. General Edward Lawrence Logan International Airport near Boston maintains an owl control team to deal with birds that constitute a threat to air traffic. The "wisdom" of owls notwithstanding, these Arctic breeders seem unable to safely gauge the comings and goings of aircraft and so constitute a hazard to landing and especially departing jets. A four-pound bird ingested by a jet engine is bound to complicate a takeoff.

Sitting upright or crouched, Snowy Owls are easily dismissed as lumps of snow, until their swiveling heads turn your way and the birds pin their yellow, cat-like eyes upon you. To be pinned in the gaze of a Snowy Owl is to be anointed by winter and initiate you into the privileged order of Snowy Owl observers.

While I cannot recall my encounters with every bird I've seen, I can easily remember every encounter with every Snowy Owl outside of the Arctic.

One of these encounters involved a bird wintering in rural Sussex County, New Jersey, in the winter of 1984. Hoping to impress my new boss, New Jersey Audubon director Tom Gilmore, I drove Director Gilmore to the owl's location as delineated by the birding hotline. Pulling off at the designated location, scanning the open, snow-covered farmland for half an hour, I failed to locate the bird. About the time the penetrating cold made coffee at a local diner an attractive prospect, a local resident drove up, parked nearby, and walked toward us. Anticipating a scolding for some unknown indiscretion, we were delighted when the local called out, "You fellers, looking for that owl?"

"Yes," we admitted. "Have you seen it?"

Smiling mirthfully, barely disguising her glee, the woman pointing skyward, disclosing the bird perched atop a nearby utility pole where the bird was calmly taking our measure.

I'm sure that Tom, like me, was impressed not by my bird-finding skills but by the pride exhibited by the local resident. Having a Snowy Owl in your neighborhood confers bragging rights and brings out the best in people. A bond is cemented by an encounter with one of winter's most magical emissaries. As if mail from Hogwarts arrives addressed to you, delivered by a real live Hedwig, that reads, "Congratulations, you have been accepted into the loyal order of Snowy Owl admirers. Use your enthusiasm to inspire others." But remember to respect owls and their need for privacy.

For those living south of the normal winter range of Snowy Owls, you can still keep track of the comings and goings of wintering Snowy Owls via the website Project SNOWstorm (www.projectsnowstorm.org), a Snowy Owl monitoring program that was spawned by the 2013–2014 northeastern irruption. This program uses solar-powered radio transmitters to track wild owls in real time and has led to remarkable insights into Snowy Owl behavior.

But nothing, I repeat, *nothing* beats the thrill of finding a Snowy Owl on your own. When scanning winter beaches, pay particular attention to places where the gulls are not concentrated or where you see gulls and raptors repeatedly diving. All presumed beached Clorox bottles deserve a second look.

Resist the temptation to approach owls too closely. The bird you are studying is likely a first-year bird and therefore inexperienced and malnourished. Flushing owls is considered very poor etiquette. When the bird pins its eyes on you and stares, you are too close. Your next step should be a step backwards.

Congratulations. To be pinned beneath the gaze of a Snowy Owl is like a rite of passage. Welcome to the Order of the Talon and the Feather.

Where the Food Is

Bird abundance is all about food. Birds concentrate where they find it or move on in search of it. With gulls, for example, this means a nomadic existence, with birds forever in search of the next opportune food resource—a die-off of winter-killed fish, a winter storm–related deposition of mollusks washed up on a beach, or a string of garbage trucks unloading at the local landfill. Opportunistic and omnivorous, if there is a meal to be had, some gull will have it. In recent years, the nimble and mostly ground-foraging Ring-billed Gull has adjusted its wintering habits to take advantage of America's burgeoning fast-food culture. Where once these inland-breeding gulls headed to ocean coasts to spend the winter, now many Ring-billeds stop short, inland, roosting on frozen reservoirs at night and fanning out during the day to exploit the gastronomic riches of America's fast-food, feasting upon items tossed by motorists at shopping center parking lots. Scanning from streetlamps, they defend their territories from rival gulls. When your discarded super-sized pack of fries hits the ground, it is theirs. Wrapper and all. Gulls aren't picky eaters.

Peaks in rodent cycles are certain to attract large numbers of raptors. Several years ago, while out caribou hunting on the south slope of the Brooks Range of Alaska, I found the mosses and matted grass of the forest-tundra ecotone nearly crawling with yellow-cheeked (or taiga) voles. So numerous were the hamster-sized rodents that when I was seated with my back against a spruce, the animals would scramble over my out-stretched legs. Nearly every spruce in view was topped by a hunting Northern Hawk Owl, and while my personal hunting success was disappointingly low, that of the owls was impressively high. The fearless owls would even perch atop the tree that served as my backrest and drop upon voles at my feet.

Also drawn to the rodent bloom were a pair of Arctic Great Horned Owls. Our resident hunting guide, Heimo Korth, con-fided that it was the first time in many years that the frost-colored birds had settled near his remote cabin. Like trappers rotating their trap lines, the owls rotated within their huge territory, set-tling each winter where they found an abundance of prey. This year the bounty was clearly in the broken forest habitat sur-rounding Heimo Korth's cabin. Boreal and Great Gray Owls, too, were concentrated in the region, and we were serenaded nightly by these doughty, far northern resident species, whose hunting success was near assured by the super abundance of prey—a boon to birds used to living in a region where winter privation is the norm. The adage "feast or famine" aptly charac-terizes the Arctic environment. And it greatly determines the comings and goings of those creatures specialized to live there.

Common Ravens are among those few land birds able to survive the Arctic winter, with many ravens settling near north-ern human settlements, foraging on food scraps. The ingenious corvids have also developed a commensal relationship with wolves, leading wolves to the frozen remains of moose and caribou. Unable to penetrate the rock-hard frozen carcasses themselves, the clever birds wait for strong canine jaws to cleave the meat. The birds then accept the scraps as their tribute, caching some for later retrieval.

Another Arctic bird that relies upon mammalian assistance for its meals is the Ivory Gull, which teams up with polar bears as they navigate the sea ice in search of seals. The gulls even consume the bloody slush around kills and are reported to investigate anything red left on the ice. Eking out a living eating meat scraps, the resourceful gulls also consume polar bear scat. In like manner, magpies wintering in Reno, Nevada, feed on frozen feces deposited in dog-exercise parks. To the magpie palate, reconstituted corn beats no corn at all.

But it is the Rock Ptarmigan, the planet's northernmost resident bird species, that sets the gold standard for hardiness and resourcefulness. Foraging in the ephemeral twilight of the Arctic winter, the flocking birds consume buds, leaves, catkins, and berries in sufficient quantity to sustain them through the night, then tunnel into the snow to roost using the air-trapping properties of fluffy powder snow to buffer them from the bitter Arctic night. Winter campers use this same principle when building quinzhees (impromptu igloos). The white nonbreeding plumage of ptarmigan render them nearly invisible against the snow, defeating the hunting eyes of wolves, foxes, wolverines, and Snowy Owls.

In the absence of tree cavities, the hardy Gray-headed Chickadee (or Siberian Tit), another permanent High Arctic resident, also burrow in the snow. Other chickadee species living in more tree-rich environs typically retreat into tree cavities for the night, as do other cavity-roosting species such as owls, woodpeckers, nuthatches, and bluebirds. Some species, like bluebirds and nuthatches, roost communally, using the warmth of multiple bodies to beat back the cold. Others, like chickadees, prefer to sleep alone but during the day gather into foraging flocks.

Roost Strategy

Roosting is a key component of winter survival. While roosting birds enter a period of inactivity typically at night, a few, like the owls, forage by night and roost during the day. Roosts may be isolated or communal, on the ground, on open bays, in dense foliage, under bridges, or on buildings, but all roosting places offer the following key attributes:

- An environment safe from predators
- Protection from the elements
- Easy access and exit routes
- Proximity to food

Roosting is not the same as nesting, and only rarely are nests used for roosting. Although, our resident Carolina Wren often snugs itself into the abandoned Barn Swallow nest on our porch after we take the holiday wreath down (the wren's preferred winter enclave). Warning to travelers returning in the dark just after the holidays: Jiggle the wreath before opening the door. Wrens are nimble, furtive, and difficult to corner.

For roosting, many smaller birds favor dense concealing vegetation or stands of marsh reed, which the birds access vertically. Multiple species find conifers particularly attractive roost sites because evergreens hold their needles year-round, offering concealment and superior weather protection. Owls select in favor of conifers, with Long-eared Owls partial to younger stands of trees ten to fifteen feet high that offer a warming southern exposure and easy access from above. A small clearing within a dense stand of natural or planted pines seems ideal. The owls are semi-communal with individual birds snugged up close to tree trunks on horizontal branches, typically facing the sun. When pines mature, creating more space between branches and easier access for predators like American Goshawk, the owls usually relocate to tighter, more protective confines elsewhere. In our New Jersey region, dense bayberry thickets are particularly prized as roost sites, but Long-eareds may choose to roost in isolated red cedars in open marsh or on the tree-impoverished prairies in mature deciduous trees. Standing erect and snugged up close to tree trunks, ear tufts raised, it takes a discerning eye to detect the elongated owls, whose orange eyes will most certainly be pinned upon you.

While prime perches are used almost every night, it is less certain that they are used by the same bird. The tiny Northern Saw-whet Owls have favorite perches as evidenced by the puddle of white droppings caked below, but careful monitoring has disclosed that different birds may use favored perches on consecutive nights. For Saw-whets, preferred perches appear to be branches three to six feet above the ground that offer easy horizontal access and typically a domed vegetative canopy for concealment and weather protection. The heads of roosting birds often touch the canopy, and honeysuckle tangles are particularly favored.

But some bird species eschew concealment entirely, preferring, instead, a buffering natural barrier to ensure protection from predators. Many waterfowl species roost in "rafts" of birds on open water far from land. Roosting Sandhill Cranes maintain security by standing collectively in shallow rivers and on sandbars, beyond the reach of coyotes patrolling the riverbanks. American Robins, which roost by the thousands in the white cedar stands near our South Jersey home, may also snug into greenbrier or phragmites tangles, whose latticework of thorns deter predators. The birds may even roost on the ground in the trackless *Spartina patens* grass on open tidal marsh lying above the reach of the tide.

Wherever birds roost, the location must be within easy commuting distance of food or prey. I recall once being shown a Bald Eagle roost snugged within an elevated mountain canyon near Salt Lake City. Every afternoon, the birds would fly back up canyon; but in the morning, the birds would use gravity to glide down to the waterfowl riches of the lake or to patrol for dead jackrabbits along roadsides. Game birds choose to roost in elevated vegetation or on the ground. My California in-laws have a world-class rose bush, the size of a UPS delivery van, named Mer, and every evening 30 to 110 California Quail rocket deep into Mer's protective confines and remain until dawn. Wild Turkeys spend their days foraging on the forest floor but at night seek elevated perches beyond the leap of coyotes.

The winsome and widespread Northern Bobwhite quail prefers to circle the wagons at night, with flock members standing shoulder to shoulder in a closed circle, all tails in, all eyes forward. The ideal quorum appears to be nine birds, with larger flocks apportioning themselves into multiple circles.

Some birds seek shelter in natural or excavated tree cavities, which afford both protection and an added layer of warmth. These include chickadees, nuthatches, bluebirds, and multiple owl species. A few birds have adopted man-made structures like barns, bridges, and residential buildings. Among them are the ubiquitous House Sparrow that finds ivy-covered walls irresistible, as do Rock Pigeons and starlings, whose mesmerizing pre-roost aggregations and fluid wheeling flight (or "murmurations") invite the interest of poets and scientists alike.

Birds may also begin going to roost by the early afternoon. Among species that roost socially, like blackbirds and crows, early arrivals get the prime perches. The most intriguing roost site I ever found was a shallow grotto on a south-facing rock cliff overlooking Alaska's Colville River. The bottom of the cave was covered in several inches of regurgitated raven pellets (the compacted and cast up undigested remains of prey). Visiting in summer, I could not know whether more than one raven was taking shelter during the long Arctic night; I can only say that the accumulation of pellets attested to a prolonged tenancy. Clever birds!

A biologist friend in Alaska advises that, to stretch out their shortened winter foraging hours and to conserve energy, ravens in Anchorage use the heat-plume venting from the city's power plant to climb aloft at dusk before setting their wings and heading to roost.

But no matter how birds roost, or where, they have one thing in common: all birds want to face the winter night with a full crop. As important to their survival as their heat-trapping feathered armor, a full crop is what keeps birds warm until morning. This is precisely why birds swarm to your bird feeders just before dark and why it is important that you keep feeders topped up with seed late in the day. A backup suet cake or seed cake will be welcome if circumstances prevent you from attending feeders before dark. Suet is energy packed, relished by many species and sold widely. Just remember to bring suet blocks inside in the evening. Suet is relished by raccoons, too, and where they occur, black bears (at least until they hibernate). When I was a child and before store-bought suet cakes were widely available, I used to entreat beef-fat trimmings from our local butcher. Stuffed into a paper bag, the challenge was getting the fat home before the grease-sodden bag tore. The butcher never charged me and, with the trimmings crammed into a homemade wire-cloth feeder, my Downy Woodpeckers relished the offering. Good luck finding that friendly neighborhood butcher today.

Strength in Numbers

In winter, many birds become flocking species, among them blackbirds, robins, and waterfowl. In parts of the South, blackbird roosts can be so large that they assume plague proportions. And while many people regard the American Robin as a harbinger of spring, these portly thrushes winter widely all across much of North America and coastally north to southern Alaska and Newfoundland. People don't commonly see robins in winter, because the birds have switched over from earthworms to a diet of fruits. Robins won't return to your suburban yard until spring rains soften the soil, bringing earthworms back within reach of robin bills. Here in South Jersey, there is an expansive sod farm strategically located beneath the main flight path for robins heading to and from their roosts in the holly forests around Millville (nicknamed "The Holly City"). Following a February shower, the rain-sodden grass is festooned with strutting robins— a kaleidoscope of hundreds of brick-breasted birds striving to supplement their winter diet.

Waterfowl concentrate where they find open water and grain fields. It used to be that the marshes of Delaware Bay hosted tens of thousands of wintering Snow Geese, drawn to the wealth of easily exhumed *Spartina* grass roots made pluckable by the mud-softening daily tides. In recent years, reduced snowpack farther north has induced many geese to winter north of our coastal marshes where the now snow-free North Jersey land-scape has unveiled a sustaining bounty of waste corn and soybeans scattered across the agricultural fields of "the Garden State"—a windfall for hungry geese.

Flocking is also a means of collective security. Many eyes see approaching danger better than two. And large bird flocks are certain to draw hunting eyes. Geese and cranes are hunted by eagles, shorebirds and Horned Larks by falcons and harriers. If targeted by a hunting raptor, shorebirds seek the safety of the sky, coalescing into a predator-defeating ball of birds whose wheel-ing flight appears like smoke signals on the horizon—a message flashed to other shorebirds that reads: "Heads up, hungry falcon on the prowl." Climbing higher and higher, either the flock outlasts the hunter or sacrifices a weaker member, ensuring the survival of the rest.

For waterfowl, flocking also helps keep open water from freezing. In the mid-Atlantic states, the introduced Mute Swans serve as icebreakers, their large bodies churning through the ice laid down during the overnight hours, creating stretches of open water that permit smaller fowl to reach down for aquatic vegetation. Coot may even pilfer vegetation from the bills of swans, who appear to genially accept a nominal amount of shrinkage from their spunky vassals.

The Amazing Survival Strategy of the Spectacled Eider

One of the most impressive examples of winter flocking and ice maintenance is found among Spectacled Eider, a mostly coastal breeder in Siberia and Alaska that, after the breeding season, masses by the thousands in open polynyas in the ice-covered Bering Sea between Alaska and Siberia southwest of St. Lawrence Island. Here, an estimated 333,000 eiders sit out the Arctic winter, with aggregations of up to 10,000 birds. While the polynyas themselves are formed by warmer, saltier vertical water currents emanating from the sea depths as well as strong parallel-flowing ocean currents, it is not unreasonable to presume that the cumulative body heat and churning action of wintering eider helps keep surface water from refreezing during particularly cold stretches. Concentrating where they find mollusk-rich beds covering the sea floor, the hardy diving ducks plunge 30 to 70 meters in search of shellfish and other invertebrates. An amazing strategy that flies in the face of the harsh northern environment.

While air temperatures in this region often fall to $-50°F$, water temperatures are 26°F warmer and the ducks are encased in filament-rich eider down, the warmest natural substance known to humankind. Out on the ice, the ducks are beyond the reach of most predators. This security coupled with the abundance of food makes for a successful wintering strategy, extreme as it may seem. Bracketing sea ice allows birds to haul out between bouts of feeding and appears to be an important component of prime wintering locations.

44

Shorebirds, too, cluster for warmth with birds on the outer rim of the flock blocking the heat-leaching wind for interior members. The intrepid Rock Sandpiper, which breeds along the shores of the Bering Sea, winters as far north as Cook Inlet, Alaska. To meet the energy demands imposed by the bitter air temperatures, the birds must forage at every falling tide for tiny clams before the mud freezes. At this latitude, in December and January, this means birds must often feed in the dark and rely upon the scouring action of ice-borne tides to expose mud soft enough for birds to penetrate with their bills.

While most shorebird species winter well south of winter's reach, along the Atlantic Coast the Purple Sandpiper, a close relative to Rock Sandpiper, winters as far north as southern Greenland. They find there a dependable supply of mussels that festoon the rocky shores and, so, confer upon Purple Sandpipers the title of our "northernmost wintering shorebird." In Europe these hardy sandpipers winter even as far north as Franz Josef Land in the Arctic Sea. At 80 to 82 degrees north latitude, it's the northernmost island group in Eurasia. For reference, the Arctic Circle lies at approximately 66 degrees 33 minutes north latitude.

Further Remarkable Strategies

The commensal relationship between ravens and wolves has already been noted, but the widespread American Crow, another communal rooster, uses its roosting strategy to communicate the whereabouts of food to flock-mates. After returning to roost at dusk, unsuccessful flock members note the bulging crops of more fortunate flock members, then follow these successful foragers out the following morning. Where one bird finds food, others may profit.

Even where food is abundant, limited hours of daylight may limit a wintering species' northern range. Turkey Vultures that once fled northern New Jersey in winter have, in recent years, expanded their winter range north to the New York / New Jersey border to avail themselves of the surfeit of road-killed white-tailed deer, whose carcasses dot roadsides every morning. The thermal-dependent scavengers were, initially, challenged by winter's shortened hours of sunlight, which limits the strength and duration of the thermals that vultures depend upon for lift. The strategy vultures devised was to roost on hilltops overlooking interstates, then use gravity to glide to the highway at first light to collect their tribute of road-killed deer. Later, the food-burdened birds use the strong midday thermals to gain enough altitude to return to roost, to be in position for the next day's foray. Their crops bulging with venison, the birds sit out the night with enough food to survive even the coldest winters, which at this latitude often fall below 0°F.

As mentioned, many birds have colorful breeding plumage but molt into more cryptic garb outside the breeding season. An exception is the Blue Jay, which maintains its electric blue plumage year-round. Even as a child, I marveled that the birds got away with this. In the stark black and white winter landscape, the electric blue birds are as conspicuous as a flare on a moonless night, easy prey for hunting eyes. One day while deer hunting, I studied a troop of jays foraging across the winter landscape. Numbering seven birds, moving as a coordinated pack, they would form a defensive perimeter with sentinels taking strategic perches, watching for danger. This strategic array allowed two members of the flock at a time to move forward and forage for acorns on the forest floor where jays are especially vulnerable but were, now, protected by the flock's sentinels. The flock's discipline was exacting; never did I see more than two birds in motion at once. But once, I saw a sentinel shift the location where it perched within a beech tree that was still holding its leaves. Even partially hidden, the bird's bright blue plumage blazed through the concealing foliage, disclosing its location to me and the other jays. Then it hit me. The jay's bright plumage bolstered situational awareness. At a glance, every member of the troop could pinpoint the precise location of every other member and know that it was protected from that quarter. As the flock moved through the forest, the protective formation of birds moved with it. It was the perfect offensive and defensive strategy aided, rather than compromised, by the bird's bright color. If the United States military wants to improve troop movement across hostile territory, I urge them to study the Blue Jay's foraging tactics. Although I stop short of suggesting troops don brightly colored uniforms. British soldiers tried this tactic during the Revolutionary war with unfortunate results.

Another species that excels at predator awareness is the Black-capped Chickadee, whose flock members not only maintain constant communication with other flock members but neighboring flocks as well. The approach of a hunting hawk is relayed through woodlands, flock to flock, via the number, pattern, and volume of *dee* calls in the classic *chick-a-dee-dee-dee* sequence. The coded sequence denotes the nature of the danger as well as the level of danger. Adding *dee* calls to the sequence denotes an elevated level of danger—the rapid approach of a Northern Pygmy-Owl as opposed to, say, the presence of an approaching tabby. A high-pitched *zee* call means imminent danger, "freeze." Later, a conversational chickadee call given by a senior member of the flock signals "all clear."

Many mammals, including black bears and woodchucks, can store enough fat to enable them to sleep through the winter. Most birds, whose metabolic needs are greater, don't have this luxury. In general, birds forage in daylight and sleep at night, conserving energy. But some birds, like chickadees and hummingbirds, can enter a state of torpor on especially cold nights, lowering their body temperatures and metabolic demands to conserve energy and make fat reserves last the night. Hummingbirds, whose metabolism is the highest of any bird, employ this strategy to survive cold nights at high altitudes where flowers abound but nighttime temperatures may plummet to near freezing.

But only one bird species that we are aware of hibernates. This is the Common Poorwill, a small nightjar of western environments. In 1946 and again in '47 a lone Poorwill was discovered hibernating in a rocky cleft in the Chuckwalla Mountains of southern California. Its body temperature was 64.4°F and its breathing barely discernable. Even into the nineteenth century, ornithologists postulated that swallows spend the winter hibernating in the mud at the bottom of ponds, a theory embraced by Aristotle. Today we know that swallows are migratory, heading south as insect numbers diminish, returning to North America only when warming temperatures support flying insects once again. The swallow concentrations noted by early students of birds were in response to insect activity, which on cold mornings will be concentrated just over the warmer surface of the water.

The Miracle That Is a Seed

Approximately 370 million years ago, plants developed an ingenious advance in their reproductive strategy: They began encasing their embryos in protective external shells. These proto-seeds allowed plants to spread across the terrestrial world, and approximately 250 million years later, seeds were discovered by birds to be a nutritious, abundant, and durable food resource that retained their nutritional value even into the winter months, thus providing insect-eating birds with an alternative food resource when winter temperatures deprived them of their primary food. Supported by this new dependable food resource, birds were, now, able to take full advantage of their heat-generating metabolic capacity and remain in colder northern latitudes even during the winter. In short, plant seeds changed the game, giving many northern-breeding birds the latitude to avoid the risks and energetic demands of migration and remain closer to, even within, their breeding territories year-round, even into Arctic regions. This section will explore the sexual revolution that changed the world of birds and led, ultimately, to the popular practice of backyard bird feeding—a multibillion-dollar industry that has itself altered winter bird abundance and distribution.

Seed-bearing plants are divided into two groups. Gymnosperms (literally "naked seed"), mostly have needles and encase their seeds in cones (examples include pines, junipers, spruce, and cedar). The other seed-producing plant group, the angiosperms, encase their seeds in ovaries or fruits (including grasses, sunflowers, oaks, birches, hickories, grapes, bayberry, and fruit trees). Birds consume both seed types.

Plant seeds are highly nutritious (providing approximately 150 calories per ounce) with a very high protein content as well as polyunsaturated fats, in addition to other vital vitamins and minerals. Seeds are durable, drought resistant, and typically overwinter in a dormant state, maintaining their nutritional value and germinating only when conditions are favorable. Some seeds can remain viable for up to five years. Best of all, seeds are abundant, with single plants in some species producing thousands in a growing season.

It took birds some time to catch onto this new food resource. The first seed-eating birds (a determination based upon bill shape) did not appear until about 120 million years ago, the early Cretaceous Period. So approximately 250 million years after plant seeds evolved.

Seeds range widely in size, from the dust-sized orchid seed to the coconut. Seeds in North America that are favored by wild birds range from the 1- to 2-millimeter birch seed to the 13- to 63-millimeter-long acorn. Pine nuts, a popular seed type favored by many northern forest birds, are 7 to 10 millimeters. For comparison, the average sunflower seed is 12.7 millimeters. But since that game-changing dietary discovery by birds, many plants and birds have evolved a commensal relationship, with plants

encouraging birds to forage on their fruits and bird species dist-
ributing seeds via caching and defecation. If you have ever
wondered why poison ivy spreads so quickly, look to birds. Follow the trail of droppings, which contain viable seeds, back to
the host poison ivy vine. Poison ivy berries are relished by
multiple woodpeckers, mockingbirds, Yellow-rumped Warblers,
bluebirds, and many other species during the colder months. The
plants even signal their fruiting readiness by donning red leaves.
Red is the universal plant-to-bird communication that reads:
"Hey guys, soup's on. Come and get it." It's why hummingbirds
are particularly drawn to red blossoms.

Yes, plants *want* their seeds to be eaten.

It is the abundance of seeds as much as their durability and
nutritional value that makes them such a vital food resource. The
average mature oak will produce 2,200 acorns per season. The
production of pine nuts is cyclic, varying greatly from year to
year. But pinyon pine nuts, whose nutritional value has been
compared to beefsteak, have high cone production every two to
seven years. Pinyon pines cover thirty-seven million acres in
Arizona, Colorado, New Mexico, and Utah. Individual cones
average ten to twenty seeds per cone. In good years, a single acre
of forest can produce 250 pounds of seed. And pinyon pine is
just one pine species. All conifers produce seed-bearing cones.
There are 600 conifer species on the planet, with juniper being
the most common genus of conifer in North America. Juniper
berries (modified cones produced by the female tree) are not
only an essential component of gin but relished by a host of
wintering birds, among them the American Robin, bluebirds,
chickadees, Yellow-rumped Warbler, and Sharp-tailed Grouse.
Robins and Townsend's Solitaires are reported to consume up to
200 juniper berries per day.

While some species of birds like doves and jays swallow
seeds whole, other species must dehusk the seed to access the
kernel within. Some like chickadees, titmice, and nuthatches

hammer them open; others, most notably the finches, have powerful seed-cracking bills that crush shells, allowing the bird to whisk out the exposed kernel with their tongue, swallowing the morsel whole. The bird then lets the husk fall from their mouth. Different finch species have bills calibrated for different sized seeds—a structural linkage first suggested by Charles Darwin in his study of Galapagos finches. In general, smaller-billed birds like American Tree Sparrow and American Goldfinch are more effective at husking small seeds, like gold-enrod and millet. But larger-billed species, like Evening Grosbeak and Northern Cardinal, are able to forage on larger seed types. In times of shortage, larger-billed birds have a com-petitive advantage, able to access a wider range of seed sizes, large and small. While the bills of all seed-eating birds have evolved to meet the task (or husk) at hand (or bill), the overlap-ping bills of the pine nut–consuming crossbills show, perhaps, the highest degree of evolutionary refinement. Crossbill bills differ from the bills of other finches by having curved and overlapping tips. Different Red Crossbill subspecies even have bills calibrated for the cones of different, specific pine species. By inserting the bill between the overlapping, shingle-like scales of the cone and exerting downward pressure, the unopened scale is pried apart, giving the bird access to the seed tucked within. The design of the bill is ingenious insofar as it allows crossbills the latitude to bring their stronger bill-closing muscles to bear upon the task, with the upturned tip of the lower bill now exerting scale-opening upward pressure. The large bills of the Cassia Crossbill (a newly designated species, endemic to the Cassia Mountains of southern Idaho) are per-fectly sized to access the scales of lodgepole pinecones. The Pinyon Jay, a crestless, nomadic, highly social corvid of western pinyon-juniper environs, uses its pointed, probing bill to access seeds in green (unopened) cones. This messy process forces birds to habitually wipe pine sap from their bills, but it gives the jays access to a food source beyond the reach of many other bird species. Though irrevocably tied to its namesake pine whose range neatly coincides with the jays, like all jays, the Pinyon Jay is opportunistic and omnivorous, eating a variety of seed types and animal matter. Once extracted, the jays will consume a pinyon seed on the spot or cache it for later retrieval.

Caching seeds for later retrieval is a widely practiced safeguard against times of shortage. Some caching birds may store up to 60 percent of their winter food stocks using this technique. The relocation of stored seeds requires extraordinary spatial memory capacity, a faculty facilitated among seed-caching birds by the enlarged hippocampus region of the bird's forebrain. Even so cerebrally endowed, not all hidden seeds are retrieved. Those seeds hidden early in the season are often overlooked and never recovered. Many of these forgotten seeds later germinate, facilitating the spread of the plant and insuring food for future generations of birds. It has been advanced that most eastern oaks sprout from acorns buried but never retrieved by Blue Jays. As for volume, one California homeowner estimated that his local troop of Acorn Woodpeckers had stashed 700 pounds of acorns in the walls of his home. Woodpecker granaries (storage bins) may contain 50,000 such stored nuts, festooning the walls of homes and outbuildings in a fresco of half-embedded acorns.

Rarely do birds deplete natural food stocks. Rather than exhaust a resource, birds rotate to more productive foraging sites. Among bird-eating hawks, these happy hunting grounds include your and your neighbors' backyard bird feeding stations. Bird-eating hawks ensure themselves a dependable food reserve through the winter by rotating to different favored perches that overlook primary and secondary hunting areas. The hunters fly perch to perch as success or the lack of it warrants. And while many homeowners are dismayed by hawks killing birds at their feeders, the hawks are actually performing an important and natural service: removing diseased members from the flock before their debilitating affliction can spread. In this capacity, hawks serve as the guardians, not the enemies, of your flock.

If it is any consolation, by feeding birds, you are not causing them to be killed. The hawk is going to consume two songbirds per day, no matter what. By feeding birds in your yard, you are only locating this natural dynamic where you will see it—a National Geographic Special in your own backyard.

The Great Blizzard of 1956

Between March 18 and 20, 1956, the northeastern United States was pummeled by a powerful nor'easter, a coastal snowstorm that killed 162 people and dumped 20 to 30 inches of snow in Morris County, New Jersey, where my father, mother, brothers Dave and Mike, and I rode out the storm in our two-story brick carriage house set a quarter mile back from Canfield Road, Convent Station. Drifts in excess of 14 feet were not uncommon. It was the biggest snowfall my five-year old eyes had ever witnessed, and I was not alone; it's like had not been seen in New Jersey since 1899.

What I remember most about the storm was my mother, desperate for some distraction to keep my brothers and me occupied (and her sane), throwing bread out the second-floor window onto the snow below. The result was amazing. This meager offering attracted a kaleidoscope of birds desperate for food, whose vibrancy and animation was magnified by the starkness of the landscape. In my mind I recall birds of every shape and hue, and I was captivated. The birds festooned across the snow constitutes my earliest memory, and the array of birds certainly has been exaggerated in my mind over time. In late March, with spring migration already underway, it is likely that the colorful guests included migrating male Red-winged Blackbirds, their red epaulets blazing, and (almost certainly) the troop of Blue Jays from the adjacent woodlands. As dedicated granivores, the House Sparrows that sheltered in our ivy-colored chimney were certainly present, as were crows, who seem never to miss out on a meal. Goldfinch? Likely.

Cardinals? In 1956, these crested beauties were not yet a common element of the birdscape in North Jersey's interior, nor had the House Finch yet been established. Whatever the species composition, I was captivated; having no bird field guide nor knowing the birds' names did not, in the least, diminish my sense of wonder. The nutritionally bankrupt stuff we called bread in those days may not have helped the birds from a dietary stand-point, but it did draw a crowd. Remember this was a time before backyard bird feeding was popular and our modest family budget would never have accommodated this luxury, anyway. When the bread was gone, the show ended, but the memory lingers still, my first vibrant experience with feeding birds in winter, a prac-tice I continue to this day, sans bread.

Bird Feeding: A National Pastime

One of the most significant recent impacts upon winter bird distribution and survival has to do with the proliferation of backyard bird feeders. Since the early mid-twentieth century, the number of households maintaining bird feeding stations for wild birds has exploded, with over fifty million US and Canadian households, now maintaining one or more feeders (with four to ten feeders per yard being average). The hobby of bird feeding goes back to Thoreau (1854) and American writer Mable Osgood Wright (1885), but the popular pastime proliferated after World War II as Americans began moving to the suburbs, a new hybrid, bird-rich environment boasting an array of vegeta-tion that birds find attractive, including shade trees, shrubbery, and flowering bushes like roses and forsythia.

As these suburban plants grow and mature, they turn ordi-nary suburban yards into de facto aviaries with commercial bird

feeders acting as fruiting bodies. Today, feeding birds in your yard is only slightly less suburban-centric than maintaining a fine lawn or buying Girl Scout cookies from your neighbor. And while the activity does cut into a family budget, most home-owners are able to set aside the funds needed to offer their feathered minions one or more seed types, plus suet (beef fat) for more finicky feeders (and even meal worms for wrens and mimids). Multiple bird-feeding store chains have evolved to serve the needs of the bird-feeding market, as well as garden centers and agricultural feed stores that have expanded their inventories to include bird feeders and seed. Even grocery stores and hardware stores have capitalized upon the craze, offering consumers a selection of mixed or specialty seeds, plus the specialized feeders needed to dispense them. It is estimated that a billion pounds of seed are sold annually in the US and Canada, generating $4 billion in income. This sum is twice the income generated by the annual sale of comic books.

The proliferation of feeders has prompted many once-woodland birds to relocate, fully or in part, to the suburbs. Feeders concentrate birds that would otherwise distribute themselves more widely. It may even have induced some species to winter north of their historic range. It has been suggested, for example, that a number of bird-hunting Sharp-shinned Hawks may, now, be short-stopping their southern migration in response to the concentrated abundance of prey they find farther north. With songbirds abundant in suburban New England, there is scant need for this bird-hunting raptor to migrate on to the Middle Atlantic and southern states. And while both the overall and secondary impacts of backyard bird feeding remain largely unstudied, it is clear that seed-eating birds choose to concentrate where feeders are available. And compared to bird species that are not attracted to bird feeders, the populations of feeder birds are mostly stable, while many other songbird species are declining. It is possible, too, that backyard bird feeding has facilitated the northward range expansion of once-southern seed-eating species, like the Tufted Titmouse, Northern Cardinal, and Red-bellied Woodpecker. Their ranges, once stopping short of New England, have, since the mid-century, expanded even into Maine. Certainly, climate change is this northern expansion's primary driver, but a reliable food resource would be a welcome support vehicle for pioneering species willing to leapfrog north, backyard feeder to backyard feeder.

But bird feeding's greatest impact is likely upon human practitioners, who quickly develop a fascination with their feathered minions, nudging many people toward a greater bird-consciousness, which translates into support for open space and for politicians who advocate on behalf of bird and habitat protection.

Bird feeding is supplemental to the diets of birds that will continue to draw much of their nutritional needs from natural food stocks (including those natural plantings found in your yard). And the number of birds attending feeders vary year to year, with winters marked by finch irruptions, finding feeders brimming with thistle-seed-loving redpolls and Pine Siskins, and Evening Grosbeaks, whose appetite for sunflower seeds can quickly demolish a modest seed budget and has earned the large, bulbous-billed finches the nickname "Evening Grosspig." But even in a normal (nonirruptive) year, our four South Jersey feeders serve approximately 140 individual birds per day, includ-ing the two Cooper's Hawks and lone Sharp-shinned that are attracted not to the seed but the birds coming to our feeders. I estimate that my in-law's Central Coast, California, home serves 150 birds per day, including the three Cooper's Hawks and two Merlins that target her feeder regulars.

Estimating the total number of birds being served by bird feeders across North America is complicated by the fluidity among local birds—moving from one backyard to the next. A case in point, my cross-street neighbor boasts of the eight male cardinals in her yard. Some of these, at least, are part of the set of twelve males I count in our yard. But even presuming that she and I evenly split the 140 individual birds I attract to my feeders, that still leaves 70 individual birds per household in a non-irruption year, a figure I present as a baseline for feeding stations across North America. Multiplied by 50 million bird feeding stations, this comes to an astonishing 350 million birds being served by bird feeders, a fair percentage of the estimated 7.2 billion birds residing in the United States and Canada. Project FeederWatch, a bird-monitoring effort managed by Audubon and the Cornell Lab of Ornithology, projects a more conservative estimate: 9,110,689 individuals of 585 species (a figure established in 2011). This figure combines the estimates tabulated by homeowners participating in Project FeederWatch, who monitors birds at feeders across the United States and Canada. Not intended to calculate a total number of birds coming to feeders, Project FeederWatch strives, instead, to establish a standard methodology that can be used to monitor trends in bird numbers (not overall numbers). A worthy endeavor and one deserving of your participation.

Many authorities have compiled lists of common backyard birds. While these vary region to region, Ma Nature typically apportions representatives of basic family groups according to region. Thus, wherever you live, you probably find yourself within the range of one or more jay species. If you live in the west, you may find a scrub-jay, Steller's Jay, or magpie. But east of the Rockies, the flamboyant Blue Jay is your neighborhood jay unless you live in the boreal forest, where Canada Jays dominate, or the Rio Grande Valley, where the flamboyant Green Jay is your neighborhood jay species.

No matter where you live, you can count on one or more chickadee or titmouse species to live near you. Mourning Doves and Downy Woodpeckers are near ubiquitous. Finches include the omnipresent House Finch, American Goldfinch and, in the west, the Lesser Goldfinch. Common sparrows include White-throated Sparrow, White-crowned Sparrow, and Dark-eyed Junco. And of course, the introduced but firmly established House Sparrow and European Starling occur just about everywhere we do. Significantly, all these species (and more) are drawn to backyard feeding stations. Mixing commercial seed (introduced in 1953) and other food items ensures bird numbers and diversity in your yard. And don't forget water. Birds need water to drink and bathe, even in winter.

And while this book is not intended to be a how-to guide to bird feeding, a few simple tips will help homeowners attract more birds. First, understand the importance of trees and other cover. A bird feeder stuck in the middle of an open grass-covered yard, away from woody vegetation, will attract few birds, no matter how much food you offer. Feeder birds are mostly forest species that demand the safety afforded by woody vegetation to retreat into when hawks are prowling. No trees? A brush pile near your feeder will offer birds safe haven and a place to wait for perch space to open up. A pile of brush in your yard might not be in the landscaping style of Martha Stewart, but it does attract birds and you may find the latticework of branches constitutes the most popular bird gathering mechanism in your yard.

Quality seed attracts more birds than mixes diluted with lots of cheap filler seed like cracked corn, canary seed, and milo and rape seed. Seed content is listed on the bag. What you want are mixes that offer a high percentage of black oil sunflower seed and millet. Pure black oil sunflower seed is consumed by all species and relished by many. We always offer at least one feeder dedicated solely to black oil sunflower seeds. Suet? Relished by many species. Shelled peanuts? Jays love them. Thistle? Sure to be a hit with goldfinches, and siskins treat the stuff like it was crack cocaine. If you have a nature center or bird specialty store near you, stop in and get advice about food items and feeder types that work in your region. Keep feeders filled, especially during cold spells and on cold mornings and especially before birds go to roost (mid-afternoon). Winter birds need to go to roost with crops full, so afternoons are precisely when your feeder regulars will be counting on you the most.

While birds obviously elect to concentrate at feeders, the practice is not without risk. Concentrated birds facilitate the spread of bird-related illnesses, like avian flu, a risk mitigated by hunting hawks that serve as the guardians of your flock, culling sick individuals before they can infect healthy birds. Free-ranging house cats are also drawn to feeders, but unlike hawks, house cats are not natural predators. Small felines are not native to North America and are estimated to needlessly kill 2.4 billion birds per year, many of these at feeders. Cats are capable of six-foot vertical leaps, and they kill for pleasure, not because they are hungry. Keep cats indoors. It is why they are called "house cats." And no, the bell collars don't work. Cats quickly learn to move without ringing the bell.

Predator Control

By choosing to feed birds in your yard, you have elected to have the natural world overlap with yours. Wherever there are concentrations of birds, there will inevitably be bird-catching hawks. This dynamic between predator and prey is natural and even beneficial. The hawk helps prevent the spread of disease and ensures that only the fittest individuals move their genes forward. It's Darwinism right in your own backyard.

Think of predation as a dance in which each partner moves the other toward higher levels of perfection. The songbird by avoiding capture, the hawk by refining its hunting technique.

And if it is any consolation, by feeding birds in your yard, you are not causing birds to be killed, you are only causing them to be killed where you will see the drama unfold. A Sharp-shinned or Cooper's Hawk is going to catch and consume two birds per day no matter what. These are professional bird-hunting hawks;

killing birds is what they are designed to do. Just think of the hawk as one more bird coming to your feeder, one that gets its seed secondhand, and appreciate them for what they are, superlative predators, as adept at catching birds on the wing as robins are accomplished at prying worms from your yard.

I'm not suggesting you should not care about the welfare of your feathered minions; I'm saying that our human sympathies have no standing in the natural arena. Neutrality is a hard truth and one that tries our humanity, but it is an honest truth, one I learned years ago when . . .

One winter day I chanced upon a red fox trotting along the side of a tidal creek on open marsh. Seeing that the fox was aware of my car but also indifferent, I concluded it was intent upon prey. Scanning ahead with my binoculars, I spotted an adult male Northern Harrier perched atop a muskrat house adjacent to the creek. The hawk, too, was aware of my presence but could not (I presumed) see the approaching fox. It is my practice to maintain a laissez-faire attitude with regards to the natural world. I watch and try to learn but my intrusion into the natural dynamic stops short of personal involvement.

And while I did not want to deprive the fox of a meal, the adult male Northern Harrier is my absolute favorite bird. Putting the car in neutral, I watched the drama unfold as the fox closed upon the hawk, whose head was turned mostly away. When the fox was almost within pouncing distance, it doubled back and turned up a side creek directly behind the harrier, out of my view.

The hawk, still facing the other way, seemed oblivious to its danger.

"Now or never," I concluded, and gave thought to gunning the engine but held to my convictions and just continued to watch.

Unable to see the fox, I trained my binoculars on the harrier, which suddenly and deftly lofted into the air just as the fox made its leap. By the time the fox reached the muskrat house, the harrier was well beyond reach. The fox seated atop the muskrat house looked my way, the expression on his face as good as saying, "Curses. Foiled again."

It was the perfect ending. My team won and I had not compromised my convictions by getting involved. Indeed, years later, I realized that not only was the interaction perfect, it was also likely choreographed. The hawk knew about the fox all along. Indeed, they may have played this cat and mouse game multiple times.

This insight stemmed from another interaction with a fox on another road, one bracketed on one side by woodlands and the other by open marsh.

It was my habit in those years to take my two Labrador retrievers out on a predawn walk. Obedient dogs, they heeled close and knew they were not allowed off the road. After conducting our morning walk for several weeks, one morning a red fox abruptly stepped onto the road about one hundred feet ahead. Sitting, facing us, the fox studied our approach, the very picture of calm. The dogs, whining, were eager to give chase but obediently stayed by my heels.

"Wait," I said loud enough for the fox to hear. "Wait . . . wait . . . ." and being good dogs, they did.

When we'd closed to within forty feet, I stopped and the fox stood, stretched, but stayed on the road.

"OK," I said to the dogs. "Get 'em."

In a frenzy of clicking toenails, the dogs were off. The fox calmly trotted off the road into the woods and disappeared. The dogs stopped where the fox stepped off the pavement, waiting and whining.

"Good dogs," I pronounced when I caught up. "Let's go."

This morning ritual, which the fox obviously relished, continued daily for several weeks, until the day a fox hunter ran his pack of fox hounds through the woods.

Either they killed the fox or more likely the fox concluded that enough was enough, and humans and their dogs were simply too capricious to treat with.

But my take-home was, just as it was with the harrier and the fox, there is much to be learned from nature if you have the patience and resolve to remove yourself from the equation and simply observe.

If you can adopt this discipline, you will soon come to appreciate how superb a predator the robin-sized Sharp-shinned Hawk is: a bird able to turn as fast as a mirror can mimic, and to pass through leaves and branches with the ease of smoke. But as formidable as a Sharp-shinned may be, the birds at your feeder are equally adept at avoiding capture. Their reflexes as refined as the hawk's, honed by generations of predator versus prey interaction. For hawks to be successful hunters, they rely upon the element of surprise or targeting a physically compromised target—that is, a weakened individual surrendered up by the flock to protect the rest.

Keep feeders away from plate glass windows or glass patio doors. Hawks learn to flush birds into the glass or the hawk, itself, may crash into the invisible barrier. Fifty percent of all such strikes end in fatality. If you have outside screens leave them up all winter. Screens are more flexible than glass and reduce the force of impact.

The Impact of Agriculture

While the conversion of forest and natural prairie to agricultural land has greatly altered wintering bird distribution, the impacts cut both ways, with millions of birds of multiple species routinely using agricultural land and its water and food resources to survive the winter. Ducks, geese, and cranes all feast upon waste grain left in the wake of mechanical harvesting machines. Commercial rice farms constitute an important habitat for rails, shorebirds, and other aquatic species. In arid regions, irrigated crop lands support large numbers of birds. Cover crops, grape vineyards, and almond and apple orchards all support wintering birds. Even bare, tilled farmland supports open-country songbird species like longspurs, pipits, Horned Larks, and Vesper and Savannah Sparrows. These in turn support assorted bird-hunting raptors, like Merlins, Prairie Falcons, and harriers.

The practice of spreading manure on fields attracts Horned Larks, longspurs, Snow Buntings, and pipits. These species are attracted to the insects embedded in the warm compost and any residual grain.

Special Gifts of the Season: Finding Birds in Winter

Anyone can be a fair-weather bird watcher, but winter birding has special challenges as well as special rewards. Not only are birds more concentrated in winter, but rare and unusual species are more easily found, tucked into wintering flocks like rubies and sapphires amid the garnets and amethyst. Indeed, for most of us, many bird species only occur in our region in winter.

Take the Little Gull, a delicate gull with sooty underwings that sometimes joins flocks of the larger, lankier, and much more common Bonaparte's Gull, who apportion themselves over large lakes, coastal areas, and turbulent water, like the waters below Niagara Falls and even sewage outfalls. Breeding mostly in Europe and Asia, Little Gulls do have a small breeding population near remote Hudson Bay, but these diminutive gulls winter regularly along the Atlantic coast and the southern Great Lakes.

In summer, birds maintain breeding territories, and so they are spread out over wide areas. In winter, many species—everyone from ducks to gulls to doves to House Sparrows—are concentrated, rounded up by winter into flocks, and many are driven south, offering students of birds across the United States front row seats to the northern breeders of the continent. As an additional bonus, in winter deciduous forests have shed their concealing cloak of leaves, giving forest birds fewer places to hide from our birding eyes.

Weather often determines where birds will concentrate. Every winter storm that passes reshuffles the deck, dealing students of birds a new hand, sending birds in search of less-snow-covered habitat, unfrozen water or, perhaps, a storm-tossed bounty of conch or clams on ocean beaches that attract the hungry eyes of gulls. The first challenge to winter birders is simply getting out there; prying yourself out of that comfy chair in front of the TV to embrace this most exciting of seasons.

You aren't hibernating. Put down this book. Grab binoculars and get out there. Take a walk. Every riverside bike path or old woodland carriage road is a portal to discovery. The birds are out there, busy being birds. Not evenly apportioned, of course; birds go where they find shelter and food. You want winter birds? Just add water. A dripping faucet, a puddle beside the road, a spring-fed seep snaking through a tangle of brush, the fast-flowing outfall of a hydroelectric plant, the moderate rapids in a river or stream: all of these are magnets for birds when the rest of the world turns to ice. Walking the west and northern shores of frozen lakes makes for easy hiking and puts you in touch with birds that will be drawn to the narrow band of melted water that rims sun-warmed shores. Birds need to drink and bathe even in winter, and all manners of aquatic invertebrates collect where ice turns to liquid. American Robins are compulsive bathers and cluster along melting lakeshores. Once while participating in a Christmas Bird Count on a bitterly cold December day, our party found the count's only Lesser Yellowlegs patrolling a puddle the size of a wash basin beside a sun-warmed asphalt road; it was the only open water to be found, but the bird had found it. Another winter, following a blanketing snow, an American Woodcock took to probing the bare soil beneath our dripping outside faucet, where it remained for several days.

Sewage treatment facilities with open settling ponds often remain open in winter and even support insect activity that might

attract lingering flycatchers and warblers, not to mention water-birds and gulls. Many such facilities encourage wildlife viewing.

Gearing Up for Cold

In this age of heated car seats and temperature-controlled indoor environments, we humans have lost many of the basic outdoor skills practiced by our ancestors. Winter birding is all about staying warm. You get cold, you go home. Opportunity lost. Here are some basic tips.

Like birds, you need to layer up. First comes the base layer: synthetic underwear, tops, and bottoms. It wicks moisture away from your skin and it keeps you toasty. Zip turtleneck tops keep necks warm and allow you to thermoregulate if exertion makes you overheat. Next comes pants and a top. Wool or fleece pants are warm, but I stubbornly prefer the ruggedness and comfort of jeans, maybe bolstered by a covering set of wind (or rain) pants (for bitter cold). Tops? Ain't nuthin' beats a good ol' wool sweater. Over that, a quality down jacket. It works for birds! On top of that, you'll need a generously cut shell jacket to trap warmth and protect the down without compressing it, just like the outer body feathers of birds. For your head? A stretchy wool or synthetic hat that can be pulled down over the tops of your ears works fine. Why just the tops? Because you'll want to hear. Winter birding demands attention to bird sounds—raucous jays, tapping woodpeckers, scolding chickadees, and sparrows scratching in the leaf litter.

Mittens are warmer than gloves but it's hard to focus binoculars with mittens. Try wearing both a mitten and a glove on your focusing hand. There are also mittens designed for hunting that allow your fingers to pass through a slit cut across the palm.

Feet? A critical matter, and no perfect solution exists. Stretchy Smartwool socks are hard to beat but after that it's up to you. Leather hiking boots with a robust tread excel on snowy slopes but many are less than waterproof. Don't forget knee length gaiters in deep snow. Uninsulated knee length rubber boots are cold, cold, cold. Insulated hunting or ice-fishing boots are toasty warm but heavy and cumbersome, made for standing not walking. Fabric overboots offer good waterproof, lightweight protection and a wide almost snowshoe-like base, plus a robust tread pattern. In addition, they allow you to wear your comfortable walking shoes within. Sunglasses? A must. And don't forget the sunscreen.

Have a vehicle suited and serviced for winter travel, one that will keep you out of trouble and more importantly get you out of trouble should you misjudge the terrain. Pack a snow shovel. Have a cell phone. Make sure it is charged and know that cell phone service is unreliable in many prime birding areas, including some National Wildlife Refuges. Let someone know where you are going and when you plan to return. Going on an organized birding field trip orchestrated by a local bird club will solve your logistics problems, offer safety in numbers, and let you tap the tribal wisdom of experienced birders.

Binoculars are a must. Spotting scopes are useful for distant waterfowl and distant perched raptors. Bring extra lens-cleaning cloths. For the photographer, winter brings special advantages as well as challenges. Birds are often more concentrated and the direct sunlight of a horizon-hugging sun puts a luster to all it touches, reducing shadows. But colder temperatures quickly drain batteries, so you'll want to bring spares and keep them warm.

Pack a stash of high-energy fruit bars and plan on an indoor lunch to give your body a break. Heading for Newburyport, Massachusetts, for wintering gulls? You must try the chowder (or *chowdah*). Bolivar, Texas, for wintering shorebirds: go for the chicken and sausage gumbo. Y'all want dirty rice wit' dat? Hell yes! Chicago lakefront for waterfowl: chili (of course con carne). Morro Bay, California, for wintering shorebirds? Try a completely different clam chowder served up in a sourdough bowl.

My first three-day birding excursion was a President's Day weekend trip to Newburyport, Massachusetts, back in 1977, organized by the worthies of the North Jersey Charles Urner Ornithological Club. Bitterly cold—subzero temperatures at least—I don't think our troop of birders passed a single chowder joint without stopping, and we still managed to see all the seasonal goodies—Iceland, Glaucous, and Black-headed Gulls, Barrow's Goldeneye, Northern Shrike. And (wonder and glory) a mound of snow that swiveled its head to disclose piercing yellow eyes, the great Snowy Owl, my first!

But my all-time favorite winter birding adventure was the winter I vowed to see a Pileated Woodpecker—a crow-sized woodpecker of mature forest that was just reestablishing itself in New Jersey following decades of deforestation-provoked absence. Jockey Hollow at Morristown National Historical Park, with its mature forest and network of carriage paths, seemed my best bet, and while I found lots of woodpecker workings (piled shards of chiseled wood), after a week of search and scores of miles walked, this grand woodpecker still eluded me. The problem was my search image. Being a large woodpecker, I assumed the bird would be near the tops of mature trees. Wrong! The crow-sized, chisel-billed birds like working the base of trees and fallen logs. The bird I finally saw was mantling a fallen branch on the forest floor a scant fifty feet off the trail. Not troubled by my scrutiny, the bird continued to apply hammer blows to the branch, the sound that had drawn me to the spot. I've seen scores of Pileateds since that encounter, but none so prized as the first.

Winter Means Waterfowl

For many observers, including North America's 1,130,000 waterfowl hunters, winter means ducks and geese. Every autumn, the planet's greatest duck hatchery, the Prairie Pothole Region and adjacent northern Canadian wetlands unleash a torrent of waterfowl that apportion themselves across North America wherever open water is found.

Some species, like the Green-winged Teal and Lesser Scaup, take up winter quarters as far south as Mexico; others like the hardy American Black Duck winter north to Newfoundland, and brant concentrate along both the Atlantic and Pacific Coasts. Canada Geese, the emblematic bird of the US Fish and Wildlife Service, are ubiquitous and these are your bellwethers. Where you find Canadas, you'll find other waterfowl.

Some of North America's greatest waterfowl concentrations occur in the Sacramento Valley of California, near Socorro, New Mexico, the bottomlands of Arkansas, the Great Lakes, and of course the incomparable Barnegat and Chesapeake Bays, where unbridled market gunning once decimated waterfowl numbers.

But if all you have close to home is the tiny duck pond in your local park or municipal golf course, it will still be a magnet to winter waterfowl so long as a portion of the water remains unfrozen. There you may be assured of a seasonal tribute of ducks and hours of viewing pleasure as winter visitors join ranks with the resident Mallards and domestic fowl. Herons and egrets, too, are attracted to any open water, as are coots and grebes. Expect a turnover in birds as winter progresses, so return often to see what newcomers have joined the flock.

It's called bird "watching," but the vocalizations of massed waterfowl enhance the viewing experience. From the barnyard quack of Mallards to the snort and gabble of shovelers to the piping whistles of Green-winged Teal and the comical two-note toot of American Wigeon, wintering waterfowl make the air over winter marshes ring with sound. Taking flight, the array of vocalizations emanating from a single flock of Northern Pintail strums the soul. The squeal of a female Wood Duck as she weaves a path through bottomland swamps (*ooo-eeek!*) is guaranteed to raise goosebumps (or maybe duck bumps). The two-note honk of Canada Goose is iconic, but fewer Americans are familiar with the single-note bark of Snow Goose, the laughing giggle of White-fronted Geese, or the resonate purr of brant. All are part of the winter soundscape and music to winter-weary ears deprived of bird song.

As a bonus, in winter, most adult male ducks are dressed in their courting finery, using the fall and winter months to attract mates, which the drakes will, then, follow north in the spring. Boasting an array of colorful plumages, a mixed flock of puddle ducks brandishes all the colors of the visible spectrum from the cone-cell jarring green of a drake Mallard's head to the dapper gray garb of male Gadwall and the rakishly attired drake Northern Pintail to the harlequin-patterned male Northern Shoveler and the flamboyantly attired male Wood Duck, a bird shaped like a Spanish galleon but dressed for Mardi Gras. Female ducks are typically more cryptically attired and some-times difficult to distinguish between similar species, but in winter females will most likely be feeding close to their distinc-tively plumaged mates. It is precisely in winter that the wisdom of our National Refuge system is showcased, with impound-ment pools carpeted by birds. A generation ago, following the market gunning era that decimated waterfowl populations, then decades of population-dampening drought, North Amer-ican waterfowl numbers were pauperized, a shadow of their historic and current abundance. Even in my youth, if a flock of Canada Geese flew overhead, strangers would stop on the side-walk and exchange smiles as if they'd just shared a special moment. And indeed, they had, a special gift of the season. Even today with goose numbers assuming plague proportions in some places, the sound of high-flying geese still draws my eyes skyward and spreads a smile across my face. There are few sounds in nature as evocative as the bark of migrating geese. It is a sound that stirs a longing in our souls for places that lie beyond the horizon and a pride in the natural dowry that is every citizen's birthright.

It was first period gym class; my senior classmates and I shivered in the chill November air as we scrimmaged in a game of flag football under the watchful eye of good coach Tierriault, the "Teddy Bear," whose bearing was so stern it could make freshmen pee in their gym shorts.

It was third down and long (a condition I have come to regard as a metaphor for my life).

"Down, set, hup one, hup two . . . ." Then suddenly the game was put on hold, cut short by the harmonic bark of high-flying geese, one of the planet's most magical sounds. Reflexively, offensive and defensive players took a knee and, raising our faces, marveled at the high V of birds, projecting all our youthful longing upon them. Nobody moved, nobody breathed, and the game did not resume until the last trumpeted notes were swallowed by the cobalt sky.

Freed from the birds' spell, we shifted back into position and then it hit us. We had delayed the game without sanction and the result would be pain. Four laps around the field plus push-ups, at least. Turning apprehensive eyes toward the Teddy Bear, we saw not the expected scowl but a smile creasing his Marine Corps recruitment-poster face. And was it our imagination or did we just see him raise a ham-sized paw to remove a tear from his cheek?

"Wasn't that just too purty," he pronounced.

It was all we could keep from comporting ourselves like freshmen.

———

There are at least three winter bird festivals in North America whose special focus is Snow Geese, the tundra breeders that in winter apportion themselves in pockets across the United States and northern Mexico.

Waterfowl are not limited to wetlands; grain fields are particularly attractive to multiple goose species. But our National

Wildlife Refuges, mostly built and managed with waterfowl in mind, are favored viewing locations and much of the best water-fowl viewing is from the comfortable confines of your car as you navigate the refuge's auto route. Check road conditions at the refuge headquarters before setting out, and ask about the appear-ance of any rare or unusual species among the flocks. You may be lucky and find the handsome, chestnut-headed Eurasian Wigeon tucked in among the more common and widespread American Wigeon. Tufted Ducks sometimes join Ring-necked Ducks or scaup; Barnacle Geese or Pink-footed Geese may flock with Canada Geese.

And waterfowl are not the only bird group drawn to refuges in winter. Herons and cranes are refuge regulars. Wintering rap-tors, too, are drawn to the concentrations of birds. From eagles to Rough-legged Hawks to Prairie Falcons to Merlins to harri-ers to Short-eared Owls, if wintering raptors are what you seek, head for the nearest refuge, where it is always open season for viewing birds of prey.

When ducks rise in a flurry of wings, there's an eagle in the vicinity. Count on it.

Mixed-Species Flocks

Birds, like justice and cell phone service, are not evenly appor-tioned across the planet. If you take a walk through a typical winter woodland, you may walk miles without encountering a bird, then *wham* you are enveloped in a bird blizzard—a mixed-species flock. These vocal and animate aggregations are one of winter birding's special treats, and no two flocks are alike, their composition determined by distribution and range.

Mixed-species foraging flocks are found in forest habitat all around the world, consisting, by definition, of three or more individuals of at least two species. In the tropics these itinerant flocks may number up to sixty individuals of thirty species. In

North America, a half-dozen species is more typical, and a dozen individuals about average.

The roving flocks consist of leaders and followers, are mostly insectivorous, and combine species of similar size but dissimilar foraging techniques. In North America, flocks form in October and disband in March as breeding season approaches. Our resident chickadee or titmice species commonly form the nucleus of the flock (the leaders of the pack). Among Black-capped Chickadee, this core constituency consists of an alpha pair and up to a dozen mostly unrelated juveniles drawn from nearby territories. Defending winter territories of up to forty acres, the dominant (typically) male chickadee leads his troop on a daily route that hits the best foraging places in the territory several times a day (including the feeders in your yard). Follower species that join the flock include nuthatches, creepers, kinglets, some warblers, wrens, and it seems every flock has an attending Downy Woodpecker who, owing to their more deliberate feeding behavior, seem always the last to catch up as flocks move through the forest at a rate of about 1,500 to 1,800 feet per hour. Some flock members forage on or near the ground, others on tree trunks, still others in the sub- and upper canopy. Flock leaders tend to herald from species that forage amid the outer branches and so are in the best position to detect approaching danger. One of the key drivers behind mixed flocks is the added measure of safety it confers upon flock members. This collective vigilance gives all flock members more time to feed, thus increasing overall foraging success (and winter survival rates). A wave of birds moving through the forest also increases the likelihood of locating food. Having a mix of species with different feeding habits reduces direct competition for food and may especially benefit those flock members adept at snapping up flushed insects in flight. Across the planet, flock leaders also tend to herald from the most social and vocal of species. The sophisticated inter-flock predator alert system employed by chickadees benefits all flock members. And while all flock members may not be fluent in chickadee-speak, all have at the very least learned the chickadee *zee* (freeze) warning call.

Mixed-species flocks are one of the special treats of winter birding. Birders attentive to the flock banter of chickadees can let their ears lead them to exciting encounters with who knows how many different species. In addition, the scolding notes of chickadees and other flock members may lead attentive birders to a roosting owl (another special treat of the season).

The downside to mixed-species flocks is that they attract predators like Sharp-shinned Hawks and Northern Pygmy-Owls. Hence the need for flock members to maintain constant vigilance. But the benefits of flocking clearly outweigh the risks, which is why the practice is so widespread. A boon for hungry birds and a bonus for winter birders.

Owls and Winter Owling

Owls are special no matter what the season, but winter is an especially good time to track down these hot-button birds. First, with leaves off the trees, these mostly woodland hunters are more easily seen. Then, too, in winter many regularly occurring northern owl species apportion themselves more widely across the United States and southern Canada. In winter, several owl species are also courting or establishing territories and so disclose their presence with their vocalizations. There is always the possibility that one or more far northern owl species will stage a migratory irruption in your region, involving scores of birds and rare viewing opportunities.

Owl etiquette is important to follow for the well-being of these birds. Your intense interest in wintering owls should always stop short of harassment. Flushing roosting owls forces them to waste energy and puts them at risk of predation or persecution by mobbing songbirds whose harangue might also draw

predators. Don't press them. Owls deserve privacy and respect. They quickly acclimate themselves to normal human foot-traffic patterns, but encroachment is not acceptable.

If you do locate a roosting owl, resist the temptation to disclose the location to anyone. Even though your friends will swear on their mother's grave that they will tell "no one." If you tell just one person, in one week a hundred observers will know about the bird. Repeated harassment will force the bird to leave. You have deprived yourself of a viewing opportunity, not to mention stressing a grand bird seeking only a measure of privacy.

Under no circumstances should you use playback recordings to entice owls close. In some places this is illegal, and it is always intrusive. Don't let your desire to get a thumbs up on social media cross the bounds of owl etiquette.

Likewise, do not use live bait to lure birds in for a picture. These are professional wild birds; let them live accordingly.

Owls are memorable, charismatic, and make for great winter birding stories. But even the sighting of a common resident owl species is a treat, including:

Great Horned Owl

This most common and widespread of the owls is aptly named. Great Horned is a milk jug–sized, barrel-shaped bird with sleepy eyes and the devil's own horns set atop its head. Found everywhere south of the tundra, even in the forest-tundra interface, these mostly nocturnal predators roost by day and may begin foraging even before dusk settles. It is not likely that a sizeable woodlot, tree-lined cemetery, or city park in North America does not host a pair. Roost sites include mature trees (coniferous or deciduous), cliff ledges, and topped off saguaro cactus trunks. Favored hunting habitat includes woods, fields, farmland, marshes, deserts, and small towns with a woodland component.

Despised by crows, observers are often alerted to a Great Horned's presence by the harangue of mobbing crows, whose raucous calls take on a particularly strident quality when directed upon the crow's arch enemy. Showing classic owlish orbits around the eyes and mostly barred underparts, Great Horned Owls range from grayish brown to nearly white in the Arctic, to sooty brown in the Pacific Northwest, to pale gray in the south-western deserts. The "horns" are actually feathered tufts, and are almost always visible. A Great Horned's barred underparts easily distinguish them from the slimmer streak-breasted Long-eared Owl and much smaller screech-owls.

Despite their nocturnal proclivities, in winter Great Horneds often become active an hour before sunset, with adults perching in full view at the edges of woodlands, and even sallying out to perch on exposed perches where they are bound to draw the ire of diurnal hunters like Northern Harriers or Cooper's Hawks. Male and female Great Horneds commonly call back and forth during December through February. This is the classic *Who, who, who / who; who-who-who* call, known to all and beloved by Holly-wood filmmakers. Other calls include shrieks, hisses, and barks. Great Horneds nest early (January to February), usually by appropriating the abandoned nests of another large bird species like herons, Osprey, and very typically Red-tailed Hawk, the owl's diurnal food-chain counterpart. Nests situated near the tops of tall trees on the edge of woodlands are preferred. Cliff ledges, squirrel's nests, and even old duck blinds in open marsh may also be used. Scan carefully; with humans in sight, birds typically crouch low on nests with only the wind-stirred ear tufts visible.

Screech-Owls (Eastern, Western, and Whiskered)

These small coffee can–sized tufted owls resemble somewhat miniature Great Horned Owls. Mostly a woodland species in the United States, they are scarce only in the tree-poor prairies. Most are gray, but Eastern Screech-Owls also come in a rufous/

brown morph. The birds are best known for their descending whinny call and warbled tremolo song. However Western Screech-Owls make a series of short whistles that recall a bouncing ball. These birds are most vocal in the hour before sunrise and can sometimes be coaxed close by whistled imitations of their calls, even (sometimes) in daylight. But observers are most often alerted to the owl's presence by the scolding harangue of chickadees and other small birds that crowd around roosting screech-owls or their cavities. While typically roosting deep in an abandoned flicker hole or tree cavity, on sunny afternoons, the small owls sometimes fill openings with their faces and may also roost in small conifers on horizontal limbs.

Cavity-rich apple trees appear to be particularly favored roost sites for this species. Any gnarly old apple tree you encounter is worth a look, even if no owls have been seen there before. Braying jays crowded around an opening in a tree are almost certain to lead you to an owl. Snug in its cavity, the owl may choose to simply stare the jays down, giving you the opportunity for study.

In the snow beneath your bird feeder, the feathered imprint of owl wings will often disclose a nocturnal interaction between a screech-owl and a rodent drawn to the spilled seed.

Spotted Owl
Large, tuftless, grayish to brownish, these owls indicate a mature (climax) western forest. The northwestern or "Northern" Spotted Owl seems particularly needful of old-growth coniferous forest, although southwestern birds (Mexican Spotted Owl) occupy shady riparian canyons close to cliffs that offer a mature tree component in whose limbs the birds often roost. While typically solitary, pairs may roost within sight of each other and, while quiescent, may be fairly conspicuous. Calls are a short series of hooting or barking followed by a protracted *whooah*. The Spotted Owl is an endangered species, and any harassment is unlawful.

Barred Owl

These tuftless, barrel-shaped owls of mature swamp and bottom-land in the southeast and mature coniferous forest in the north and the northern Rockies have soulful brown eyes and streaked underparts. Mostly nocturnal, during the day these birds retreat into large tree cavities but may become vocal in broad daylight if provoked by a loud noise (like a car alarm or noon-day factory whistle). Their classic vocalization is a low resonate hoot sequence, phonetically rendered *Who cooks for you? Who cooks for you? . . . Who cooks for you all?* These owls also emit an impressive array of gurgles, shrieks, and barks.

Northern Saw-whet Owl

These small, tuftless, coffee mug–sized owls with brown-streaked underparts and winsome faces breed in northern and mountain forests, and winter widely but sparingly across northern and western states south into Mexico. Roosting in an assortment of protective tangles that offer easy access and escape, they seem to favor conifers and holly trees but are also drawn to honeysuckle tangles that offer perched birds a protective dome. Saw-whets may actually perch with their heads touching the vines. Favored perches are used night after night, but not necessarily by the same owl. Over the course of the winter, these favored perches are distinguished by the accumulation of caked whitewash (droppings) on the forest floor. Perches may be only three feet above the ground but are typically higher.

Their song is a series of measured whistled toots that recall a file drawn over a saw blade. Calls include a rising cat-like screech and barks. Saw-whets have the unfortunate habit of flying at automobile-grill height, resulting in many being killed by cars.

Long-eared Owl

This medium-sized tufted owl resembles somewhat a gaunt Great Horned Owl—a Great Horned drawn by El Greco.

Widespread but common nowhere except in winter when birds may roost communally in place. Long-eareds sally out at near dark and hunt by coursing low over marshes, meadows, and fields, much like the more crepuscular and similarly sized Short-eared Owl that, in poor light, Long-eareds are easily confused. Roosting birds commonly choose a stand of ten-to-fifteen-foot-tall conifers near fields, with stands offering a sunny southern exposure preferred. Young planted pines and shelter belts are particularly attractive. Roosting birds stand rigidly upright on horizontal branches, often close to trunks. Birds are most often discovered because of the spattering of white droppings beneath favored perches and the accumulation of regurgitated owl pellets the birds cough up during the winter. Roosting from eye-level to mid-height, where you find one Long-eared, there are commonly more. When walking through a potential roost grove, avoid flushing the birds by walking slowly and studying the ground (and branches) ahead for whitewash. Stop and scan the branches above the droppings for the bird that will already be watching you. Now look left, right, and above, because you may already have blundered into a communal roost.

Calls include moaning, squealing, barks, and mewing, but in winter Long-eareds are mostly silent. Their orange facial disks are distinctive. In especially cold temperatures, the birds may fluff up, recalling Great Horned Owls, but note their mostly streaked underparts (barred on Great Horned). These birds also roost in bayberry thickets and even isolated cedar trees in or near meadows.

Short-eared Owl
Wow! An owl that is active in daylight. At dawn and dusk and on cloudy days, anyway. Tuftless and overall tawny, the birds recall large, nimble, stiff-winged moths. Becoming active just before sunset, the birds hunt over marshes and fields with a low, coursing flight punctuated by periodic perch time and wing-over drops to the ground.

Short-eareds are often seen sparring with the larger, rangier Northern Harrier. Where you find hunting harriers, you'll often find Short-eared Owls. When sparring, the owl will be the more compact, more nimble, and the higher aerialist. During such engagements, the owl may emit a snarling bark. Roosting mostly on the ground, often communally, Short-eareds are sometimes flushed during the day by harriers, and the squabble begins. Able to hover and adept at soaring, the owls are, nevertheless, typically seen coursing alone just above the grass. Perches are usually low and isolated. Short fence posts, duck blinds, and snow fencing are ideal. The birds do not respond well to squeaking (imitated mouse calls); a glance is all you'll get.

Barn Owl

Widespread across the United States, north to the border regions of Canada, these somber-eyed, monkey-faced birds are strictly nocturnal and their underparts so white they are sometimes mistaken for Snowy Owls. Coursing low over fields and marshes, the birds roost in barns, abandoned buildings, elevated and enclosed deer blinds, mine shafts, caves, under bridges, and even in gaps between stacked hay bales. They are readily attracted to properly sized artificial nest boxes erected in open habitat. The male bird often roosts several miles from the female, who tends the young exclusively. These roosting males are sometimes found in stands of conifers (often near the tops). Vocal in flight, their presence is often heralded by a loud hissing, goosebump-raising shriek. Barn Owls respond readily to imitated mouse squeaks and may even hover above you.

Boreal Owl

Unless you live in the boreal (spruce-fir) forests stretching from Labrador to Alaska, or at higher altitudes in forests in the Rockies south to northern New Mexico, you should not expect to encounter this northern, mostly resident species. Periodic irruptions do find some birds relocating to north-central and northeastern states south to Wisconsin and New York, but even where resident, the birds are secretive and almost always found in coniferous forest. Their song is a rapid muffled tooting that recalls a winnowing snipe. Even in Maine, the bird is less than annual.

Northern Hawk Owl

This medium-sized, square-faced, long-tailed northern owl is most often found keyed up on the springy tops of medium-sized spruce trees in broken boreal habitat located between Newfoundland and Alaska, north to tree line. Active in daylight, in winter the birds sometimes wander into northern states where they perch atop signs, utility poles, and other elevated objects in open country. While rare, they are close to annual winter visitors to Maine. Nearly fearless, the birds allow close approach as they scan for prey.

Great Gray Owl

Another northern owl likely to elude you is the huge Great Gray, a tuftless owl with hypnotic yellow eyes of northern coniferous forests from Alaska to Quebec and south along the Rockies and Sierras to northern California and Wyoming. Often found at the edge of forest clearings, the birds sometimes perch only a foot or two above the ground. Their breeding zone regularly expands to Minnesota and occasionally involves in irregular irruptive flights that have ferried birds as far south as Long Island (once). Even in Maine, this owl is considered a rare winter visitor.

Northern Pygmy-Owl

This tiny, near sparrow-sized, bird-hunting owl might easily be mistaken for a songbird as it speeds past, except for the entourage of scolding chickadees and titmice that are almost certain to be in the bird's wake. Tuftless and long-tailed, nothing seems to incite the ire of songbirds more than this ferocious western hunter. Vocalizations include slow single-note and double-note toots (double being more typical of southern birds). Found from southern Alaska to Mexico, the birds favor broken forests of mixed conifer with oaks and sycamores. In winter they may be found in city cemeteries with a tree-rich component, where they perch-hunt from horizontal branches in search of small birds. Most are grayish brown, but northwestern birds are more rufous.

Ferruginous Pygmy-Owl

Rare and local in southern Arizona and Texas, this small desert owl occurs in patchy riparian woods, even close to ranch houses with surrounding trees. In Arizona the birds are most often associated with saguaro cactus and mesquite. Active in daylight, in hot weather the perch-hunting birds seek shade. Their song is a rapid, even series of squeaky toots. A fearless hunter, it has been known to take down birds twice its size.

Flammulated Owl

More often heard than seen, this screech owl–sized owl of western forests breeds from British Columbia to Mexico. Found in dry woodlands, favoring ponderosa pine, the nocturnal birds roost in cavities or high in conifers by day (often hidden by foliage). Ear tufts might not be visible. Their dark eyes distinguish it from other owls.

Elf Owl

Our smallest North American owl breeds in dry, open southwestern forest and desert washes where it roosts in old woodpecker holes and becomes active and vocal at dusk. Their song is a rapid series of five to seven muffled yaps.

Snowy Owl

The Grail Bird among winter birders is undoubtedly this large circumpolar Arctic breeder whose normal winter range extends to the Canadian prairies and northern border states, with the eastern birds being annual south to Long Island. Our largest owl with a four-to-five-foot wingspan and a dense covering of feathers, it also ranks as our heaviest owl, weighing a pound more than the Great Horned Owl. A bird of open treeless habitat (often) with a hilly component. Winter territories favored by Snowy Owls include lightly trafficked beaches, airports, marshes, and open grasslands where the birds may perch atop utility poles, silos, and fence posts. Otherwise, the birds are comfortable sitting on open ground. Only adult males are wholly (or mostly) white; young birds, which dominate in wintering populations, have heavily flecked bodies that suggest dirty snow. Mostly silent in winter, the birds are crepuscular but largely quiescent during the day, becoming increasingly animate close to sunset. Their flight is swift and direct. Rodents and ducks are favored prey. Tolerant of humans but not indifferent, when this owl turns its yellow eyes on you and stares, you are close enough. Back off.

National Wildlife Refuges: National Treasures

Most Americans are familiar with our celebrated national parks, but few are aware of the extraordinary wildlife viewing opportunities offered by our network of National Wildlife Refuges. And while many parks require online reservations and long wait lines, refuges have no wait times, just drive-up viewing.

Probably America's best-kept secret, this network of wildlife-friendly sanctuaries is maintained by the US Fish and Wildlife Service—a branch of the Department of the Interior, employing over 8,000 wildlife professionals, whose primary constituents are the plants and animals under their jurisdiction. No other country has dedicated more resources toward maintaining its natural dowry, and the proof is seen on every drive around a refuge's network of roads and from strategic viewing platforms. From the smallest songbirds to the stately Whooping Crane, North America's wintering birds often select to be on a refuge.

Millions upon millions of ducks and geese annually pack into the impoundments and managed marshlands of these bird-rich enclaves. A living kaleidoscope of birds.

One of the most common constituents is the Green-winged Teal, a near-pigeon-sized duck, whose flocks explode from the water in an ascending avalanche, each wing flashing an iridescent green patch so vivid it would incite envy in an emerald. There is no green like Green-winged Teal green, and there is only one way to properly see it. Live and in person. Thousands of teal may cluster in places like Quivira National Wildlife Refuge in Kansas, Kern NWR in California, and Jamaica Bay NWR in New York City. But teal are only one of the scores of waterfowl species that cram into our refuges for the winter—from Snow Geese to Tundra Swans to Northern Pintails to Gadwalls to wigeon to Hooded Merganser, our refuges host them all.

These winter waterfowl concentrations are a natural spectacle, in league with desert wildflower blooms in California, the monarch butterfly roosts of Mexico, and the firefly spectacle of the Great Smoky Mountains.

Best of all, the nearest refuge is probably just a few hours' drive from your home—a National Geographic spectacle almost in your own backyard, showcasing the natural riches of the continent.

Our National Wildlife Refuges have concentrated and supported waterfowl and other bird species for over a century, providing quality wildlife viewing opportunities and offering life sustaining wetlands to water-dependent birds, even in times of severe drought. Established in 1903 by conservation-crusading President Theodore Roosevelt, there are now 588 refuges maintained by the US Fish and Wildlife Service covering 856 million acres of wildlife-friendly habitat. With at least one refuge in every state, there is bound to be a refuge near you serving the needs of breeding, migrating, and especially wintering bird species. These are visited by sixty-five million human visitors each year. The largest refuge is the 19,286,482.3-acre Arctic National Wildlife Refuge (a national treasure in its own right); the smallest (est. 1915), the Mille Lacs NWR in Minnesota (Hennepin Island), a mere .57 acres. More significantly, over one hundred refuges are within city limits or are adjacent to major metropolitan areas, including six in the San Francisco Bay area, three near Houston, and one, Rocky Mountain Arsenal NWR, near Denver. The 12,600-acre Jamaica Bay Refuge is located in the New York boroughs of Brooklyn and Queens, and is a critical stopover point for migrating shorebirds. This metropolitan refuge can be reached by subway plus an 18-minute walk from the Broad Channel station in Queens. Refuges differ from national parks by making wildlife and their habitat needs their primary management focus, not serving human visitors like the National Park Service.

Many states and assorted not-for-profits, too, maintain wildlife refuges. And organizations like Ducks Unlimited have land agreements with large landowners that benefit waterfowl and other wildlife.

Many National Wildlife Refuges offer a wildlife drive, and larger refuges have staffed visitor centers. There is typically a nominal entrance fee, but the purchase of a federal duck stamp will gain you unlimited access to any refuge. Cost is minimal. Purchases may be made at refuge headquarters, post offices, and online. Whether you use your duck stamp to gain access to refuges or not, every conservation-minded citizen should purchase duck stamps annually. 98 percent of duck-stamp sales go directly to the acquisition of critical wetlands, making duck stamps every citizen's most direct link to bird-habitat protection.

A Changing Winter Climate

At first glance, a warming climate might be thought to benefit wintering birds, but for several reasons this is not the case. The warming planet throws off the timing of seasonal food resources that migrating birds have come to rely upon. If oaks flower early, say in April instead of May, they will not host the pollinating insects that many warblers have come to depend upon to fuel their northbound migration.

Ice storms where once snow used to fall encase food in impenetrable armor beyond the reach of birdy bills. In addition, wet snow in the Arctic lacks the insulating property of the dry powder needed by roosting ptarmigan. A daily pattern of freezing and thawing may mean birds cannot burrow into the snow at all so are less able to withstand the energy sapping cold of the Arctic winter night.

Changes in the Arctic snowpack have also, in places, caused lemming numbers to decline and disrupted their population cycles. These declines force Arctic foxes and aerial predators (like jaegers) to redirect their hunting efforts from lemmings to nesting shorebirds to feed their young, thus impacting shorebird breeding success.

Winter-Breeding Birds

Somewhat counterintuitively, some bird species choose to nest in winter, among them are the Canada Jay, Bald Eagle, and Great Horned Owl. Some of these species may be incubating eggs even as snow falls in January. Early nesting gives adult owls the latitude to care for young throughout the summer and into fall, increasing survival rates. In this age of rising temperatures, birds also find it is easier to keep temperature-challenged chicks warm in winter than cool in summer, thus favoring an early nesting cycle.

In Florida, Bald Eagles begin nesting in late autumn precisely to avoid the terrible summer heat. Elsewhere in North America, eagles are incubating by mid-January, taking advantage of the seasonal abundance of winter-killed fish and starving or road-killed deer. It takes young eagles 116 days to fledge (Barn

Swallows only 41 days). If eaglets fledge in mid-summer, they then have several months of bountiful food resources to capitalize upon as they learn to fend for themselves. Then it's five years of eagle apprenticeship until immature birds reach adulthood.

Most remarkable is the breeding strategy of Phainopepla, a fluffy-crested southwestern species that has two breeding seasons in a single calendar year: one in the winter in the deserts where the birds forage on mistletoe, and another in spring in riparian woodlands where the birds take advantage of the seasonal berry crop plus the seasonal abundance of insects. Whether the same birds breed in both seasons is uncertain.

Helping Birds in Winter (and Year-Round)

As many readers probably know, North America's birds have suffered an overall 30 percent population decline over the last 30 years. The loss of three billion birds begins with a single bird. But the road to recovery does, too. The bird you save through your mindfulness and actions may be that turn-around bird. As cognizant citizens of planet Earth, we are not powerless to prevent bird decline, and bird loss is not someone else's problem to solve. It is, collectively, ours. If humans are the cause of bird decline, we must necessarily be the solution. There are still 7.2 billion birds in North America, and the means of protecting them lie within every individual's reach. Indeed, to fail to secure a future for birds is an abrogation of our obligation as stewards of the planet.

Bird protection begins in your own backyard. Make it a vegetatively rich environment by keeping lawn space to a minimum and planting native vegetation whose nectar and fruits attract the insects consumed by birds. Certain butterflies and their caterpillars are specific to certain native plant species. Non-native plants are showy, but most are less beneficial to birds than native species, and some introduced plant species are simply invasive, coopting resources that would otherwise support more symbiotic, native plants.

Resist the temptation to sanitize your forest understory or your tree-lined property edge. Nature likes diversity and is perfectly capable of feathering her own nest. Fallen leaves are not clutter, they are ground cover, preserving moisture in the soil and supporting a rich community of animal life. Many bird species are specialized to forage in the leaf litter, among them thrashers, towhees, thrushes, and assorted sparrows. Dead trees or limbs that do not constitute a hazard to people or property should be left standing. Many cavity-nesting birds depend upon standing dead timber. In our summer retreat in Maine, two dead birch trees loom over the parking area. Every spring, the landscaper lobbies to take them down, but as evidenced by the array of excavated cavities in the trunks, our Hairy and Downy Woodpeckers love them as they stand. Our troop of resident chickadees nest and roost in the vacant cavities. The solution? During threatening weather, we park our cars elsewhere. Eventually the trees will fall, decompose, and be recycled back into the forest.

At our home in New Jersey, we maintain a classically landscaped postage stamp-sized front yard for the sake of our neighbors, but our backyard, surrounded by a stockade fence, is where we and the birds live. The back lawn is just large enough for two closely monitored spaniels to pee and poop. The southeast corner of the property, once a rose garden, has been allowed to go native, bolstered by planted, native fruiting trees like wild cherry, dogwood, and crab apple, and replete with volunteer shrubs, vines, and grasses (aka weeds) whose seeds were deposited in the droppings of the birds coming to our feeders. In the tangled confines of our "nature's corner" are Virginia creeper, poison ivy, and goldenrod. It is the favored staging area for all birds coming to our feeders, a perch-filled labyrinth that keeps birds safe and comfortable as they wait for feeder space to open up. My California in-laws maintain a lush "garden area," replete with paving stone paths and multiple water sources. The canopy of apple trees provides shade, and the planted beds offer food and cover for multiple bird species, most notably Fox, White-crowned and Golden-crowned Sparrows, plus California Thrashers and Towhees, and the hordes of Yellow-rumped Warblers and Ruby-crowned Kinglets that come in for a drink of water after a day spent foraging among the vines in the adjacent vineyard. At night mule deer come in to quench their thirst. A bird-friendly yard need not be cluttered or unkempt, but native plants serve birds best.

By far the most beneficial thing a homeowner can do for birds is to not allow cats to roam. Free-ranging cats are bird-killing machines and, like kudzu, gypsy moths, snakehead (fish), Burmese pythons and feral pigs, feral cats constitute an invasive species that kill an estimated 2.4 billion birds a year. A tragic and preventable loss. Don't think for a moment that your aging, overweight, well-fed, bell-wearing tabby is not capable of killing birds. They can, and they will if allowed to roam outside. Keep cats (and your bird-killing spaniels) indoors or leashed and encourage your neighbors to do the same. This action will ensure pets longer, healthier lives and save countless birds. If you find yourself no longer able or willing to care for a cat, do not dump it in a cat colony, or at the end of some dirt road; turn it over to an animal rescue organization.

Feral cats are a menace to wildlife.

The second biggest human-related threat to birds is large plate-glass windows and patio doors. It is estimated that 50 percent of all bird strikes result in fatality. An estimated one billion birds are killed by window strikes every year. Keeping outside screens up all year reduces the force of impact. When building new homes on ridge tops or within forest corridors, design to minimize windows that lie perpendicular to (that is, across) the flight path of birds. Also consider burying your utility lines. Elevated wires seine millions of migrating birds out of the air, resulting in needless and preventable loss.

Automobiles, too, constitute a menace to birds, killing millions of birds annually. When driving at dawn and dusk, the times birds are especially mobile, be extra vigilant. If you see birds foraging on the side of the road ahead, slow down. Many bird strikes are preventable. Never throw food from car windows. It attracts rodents, which attract owls that fly at car-grill height. Ping!

Most bird strikes occur in narrow woodland corridors where birds must cross roads to reach the woods on the opposing side. Where you see deer crossing signs, slow down; these are bird crossing zones, too. And while striking a Hermit Thrush may not put your car in the shop, it still results in the senseless loss of a valuable member of the forest community. And it's one less bird to return north in the spring.

Wintering birds gather in green belts and along vegetated water courses. If these natural areas and walking paths are managed by your local park commission, lobby that they be treated not as picnic or recreational areas but as natural areas. Leave the understory intact.

If your municipality has an environmental commission or planning board, be certain bird protection is an integral part of the planning process. Fragmenting habitat renders it unsuitable for many woodland and even grassland bird species and results in changes to the forest ecology, drying forest interiors and facilitating access by invasive species. Roads and utility line cuts parcel habitats, rendering them less attractive to multiple bird species.

Support politicians who advocate on behalf of the environment, open space, and keeping federal and state wildlife agencies well-staffed and well-funded. These wildlife professionals are the guardians of our natural resources.

Consider joining a local bird-watching group to learn about wildlife-viewing opportunities in your area. Knowing when to visit a natural area is just as important as knowing where. Experienced students of birds can provide this insight and lead you to cherished encounters with birds in your area. If your local paper has a nature column, field trip dates are often posted there.

SELECTED SPECIES PROFILES OF
NORTH AMERICAN
WINTER BIRDS

Champions of the Flyways:
Birds That Go the Distance

While this book's primary focus is those birds specialized to surmount the northern winter, it would be remiss not to showcase those species whose survival strategy prompts them go to extraordinary lengths (or distances) to avoid winter. Among these champion migrants are the Arctic Tern, Northern Wheatear, and Bar-tailed Godwit.

Arctic Tern

Breeding on gravel shores across much of the Arctic, this slender winged, silver-backed seabird might more accurately be classified as an air-dweller, or perhaps sunbird, spending their brief breeding season in the Arctic and the remaining nonbreeding days in the Antarctic. Likely no other bird on Earth is bathed by so much summer sunlight as the Arctic Tern. The shortest distance between the Arctic and Antarctic is 18,641 miles. But in its search for optimal flying and feeding conditions, these four-ounce birds may travel 44,000 to 59,000 miles in their annual round trip peregrinations to and from the opposite ends of the

Earth. No species goes to greater lengths to avoid winter's reach. The oldest Arctic Tern, a bird banded in Maine, lived thirty-four years. Presuming an annual round trip migration of 45,000 miles, the bird would have flown 720,000 miles over the course of its life or the equivalent of six trips to the moon.

While summer in the Antarctic is hardly anyone's idea of a tropical getaway, with icebergs bobbing and noonday temperatures averaging just above freezing, it beats the December weather in Pond Inlet, Nunavut, which as winter darkness closes in rarely climbs above 0°F. Wintering in the Antarctic also allows these sight-guided fishing birds the entire twenty-four hours a day to feed.

Northern Wheatear
This bluebird-sized, circumpolar breeder seems wedded to boulders. In North America from western Alaska to Labrador and Greenland, they nest beneath or between them; on the plains of Central and Northern Africa, they perch-hunt from them. Hunters of the arid African plains, wheatears are one of the few North American breeders that do not winter in the Western Hemisphere, electing, instead, to adhere to the wintering pattern pioneered by their Old World ancestors.

The name wheatear comes from the Old English "white ass," an obvious reference to the ground-dwelling, thrush-like bird's snowy rump. Most remarkable about North American wheatears is their dual migration routes. Departing from the North American Arctic, flying to Africa is as about the same distance whether your route takes you east or west. While the bird's destination remains essentially the same (sub-Saharan Africa), those birds breeding in the Canadian Archipelago migrate south and east across Europe. Those birds breeding in Alaska head west, moving in stages across Asia, a journey which necessitates crossing the Arabian Desert and navigates 9,000 miles over one third of the Earth's circumference.

It's a long way for a tiny bird to travel but at least it can be done in stages, unlike the southbound flight of Bar-tailed Godwit, the planet's undisputed nonstop-migration champion.

Bar-tailed Godwit

Most of North America's arctic shorebirds are long distance migrants, but the Bar-tailed Godwit, which breeds along Alaska's west coast, is the unrivaled super achiever, regularly flying from Alaska to New Zealand, nonstop, across open water. To accomplish this remarkable feat, the large (16-inch, 12-ounce) sword-billed shorebirds spend several weeks fattening up in the bivalve-rich Yukon–Kuskokwim Delta of southeast Alaska then set off in late autumn across the open Pacific.

On October 13, 2022, a juvenile Bar-tailed banded as a nestling near Nome set off from the Y–K wearing a five-gram transmitter, and eleven days later B-6, as the bird was designated, touched down in Tasmania, a record setting distance of 8,425 nonstop miles. Flying day and night the five-month-old bird passed west of Hawaii and overflew the Marshall and Gilbert Islands, the only land along her route. Unlike phalaropes, godwits cannot land on water for more than a few moments before their feathers become waterlogged, so over most of the bird's route it was fly or die, one wingbeat after another. Perhaps most remarkable is the fact that the young bird navigated the journey without adult guidance insofar as adult godwits (like most shorebirds) depart before the juveniles.

Another long-distance champion worth highlighting is the closely related Hudsonian Godwit, which breeds in two separate populations—western Alaska and northern Canada. In June and July they relocate to James Bay, Canada, to fatten up for their 4,000-mile flight out over the Atlantic Ocean or through the Central United States to southern South America, most notably the island of Chiloé, Chile, arriving there in September and October. In spring (April and May), the birds jump off for their 6,000-mile nonstop return trip to breeding grounds with an interim stop on the Texas Gulf Coast or the Prairie Pothole Region of North America. Once slaughtered by the thousands by market gunners who called the large meaty birds the "ring-tailed marlin," the current population is estimated at 77,000 adults.

Profiles in Fortitude

The hardy species showcased in this section are, in some cases, species that have already been touched upon in these pages but whose specialized adaptations are worthy of deeper exploration. These are the extra-tough winter survivalists.

American Black Duck

This hardiest of dabbling ducks winters farther north than any other dabbler, with a hardiness supported by the duck's superior body mass—the highest mean body mass in the *Anas* genus. Resembling a dark female mallard (with which Black Ducks freely hybridize), Black Ducks' varied diet, which includes plants, snails, and mussels, supports a wintering range that encompass much of the eastern United States and extends north to Newfoundland.

The duck's partiality to salt water allows it to winter in northern regions where fresh water is mostly locked in ice. Eschewing the large flocks favored by many waterfowl, Black Ducks typically sit out the day in twos and threes in salt marsh ponds and fan out across tidal mudflats to feed. This aversion to flocks, coupled with their extreme wariness, largely buffered Black Ducks from the decimation wrought upon waterfowl during the market gunning era of the late 1800s and early 1900s. Roosting at night on open bays, the birds head inland at dawn to forage, flying just above the reach of waterfowler's guns.

Like other waterfowl, Black Ducks are designed with an ingenious "counter current" heat-exchange system in their legs, with warm arterial blood passing close to cooler blood returning to the heart via the veins, warming it. The blood flowing to the feet is necessarily cooled but still warm enough to prevent frost bite, and the result of the exchange is a net reduction in overall heat loss. The bright red legs found on some adult birds precipitated something of an ornithological Donnybrook early in the twentieth century among those who believed that these red-legged birds constituted a larger, hardier northern "form" of Black Duck, *Anas rubripes obscurus*, which would appear in late winter among duller-legged native breeders. The myth of the "red-legged" Black Duck has since been debunked, but some waterfowlers still cling to the romantic notion of the red-legged Black Duck—a duck as tough as those hardy souls willing to hunt in weather conditions challenging enough to daunt an eider.

Harlequin Duck

Everyone who sees a picture of this dramatically marked bird aspires to see one in person. But unless seen at close range, the result will likely be disappointing. At a distance, the male's bold pattern, unaccountably, homogenizes into bland uniformity, rendering the bird's bold clown-like pattern as unremarkable as the cryptically gray females.

Highly social but mostly occurring in small single-species groups, these ducks of rapidly flowing streams and churning surf lead rough and tumble lives. North American Harlequins are divided into two populations, one breeding from Greenland to northern Quebec and eastern Canada, and a Pacific population breeding on swift-flowing inland streams from Alaska to Oregon and Wyoming. After breeding, both populations gather into small groups and travel to their respective coastal habitats, close to breeding areas in the east—meaning from Greenland and Newfoundland to New Jersey—and in the west from the Aleutians to northern California. Favored habitats are rocky coasts near fast-flowing inlets and seaward-jutting headlands. Here the birds dive mostly for mollusks in the swift-flowing currents and churning waves. Built for rough conditions, the compact, short-necked birds have a remarkably tight outer layer of feathers that traps air so efficiently that surfacing birds bob like chunks of ice in the chop. Frequently dashed against rocks by wave action, many adult birds show healed bone fractures.

Roosting well offshore, on warmer days the birds may haul out on coastal rocks to preen and soak up the sun. Surprisingly vocal, the birds emit a snorting chortle and yipping squeaks, the source of the bird's nickname, the "sea mouse."

Rock Ptarmigan

OK, if you're looking for North America's toughest winter bird, look no further. This grouse of High Arctic regions winters as far as 75 degrees north, a latitude that experiences three months of total darkness in winter and temperatures as low as −70°F. It is the far-northern title holder and, very appropriately, the Rock Ptarmigan is the provincial bird of Nunavut, Canada, where the Indigenous people call it *Aqiggiq*.

Rock Ptarmigan breed as far north as summer advances and winter farther north than any other bird species, even to winter-tempered Ellesmere Island and northern Greenland. And while some extreme northern-breeding birds may relocate closer to tree line in winter, over most of their range, Rock Ptarmigan are resident and remain year-round in barren, rocky Arctic tundra where mature trees reach ankle height. Smaller and somewhat grayer than Willow Ptarmigan, the Rock Ptarmigan favors bleaker, higher, drier, rocky, hilly, sparsely vegetated tundra. Where the ranges of the two ptarmigan species overlap, Willow Ptarmigan gather in wetter areas with emergent (knee- to waist-high) willows.

Admirably camouflaged no matter the season, in summer female Rock Ptarmigan show brown and blackish upper parts and black-barred underparts. So well camouflaged are the birds that predators have been observed to walk right past incubating females, dismissing them as rocks. Males are similar to females but in summer retain mostly white underparts, and both sexes are overall colder brown than the more rufous-tinged Willow Ptarmigan. In winter, both sexes are wholly white except for black outer tails (and the male's distinctive black eye stripe).

They are superbly suited to survive the Arctic winter. In addition to their snow-colored alternate plumage and thick underlayer of down feathers, the extra-long contour (body) feathers are up to 78 percent fluff. These filament-rich "after feathers" give Rock Ptarmigan an extra layer of warmth. In addition, the feet of Rock Ptarmigan are feathered to the toes, increasing their surface area by as much as 400 percent and turning the grouse's feet into miniature snowshoes that permit birds to maneuver easily across deep powder snow. The birds subcutaneous winter fat layer is also sometimes cited as an important source of insulation. In winter the birds gather in sex segregated flocks of 20 to 250 individuals that forage for the buds and catkins of ground-hugging Arctic plants, most notably dwarf birch and white heather, with willow buds and catkins rounding out the winter diet. At these latitudes the too-brief growing season forces woody plants to abandon their elevated pride and creep along the ground like vines. When foraging in winter, the birds seek out wind-scoured hilltops where the vegetation is exposed, but birds also scratch for food beneath the snow. Indigenous people observe that Rock Ptarmigan also forage in the snow-breaking wake of caribou and musk oxen. The birds bathe in snow and, after feeding, retreat into the snow in burrows that they excavate with their feet. Thus, insulated in their snow caves, the birds weather the bitter cold and the prolonged darkness of an Arctic winter.

Rock Ptarmigan are an important food resource for Indigenous people in winter and early spring, when other harvestable animals are in short supply. The birds may be shot or snared. The world population is estimated at eight million adults, and the annual harvest is approximately one million birds.

It would be unseemly not to showcase the other two Ptarmigan species, whose hardiness approaches that of Rock Ptarmigan. The second species is the larger Willow Ptarmigan, a resident mostly of subarctic regions whose range in winter extends south even into the boreal forest. Several have been recorded in Maine.

Our third ptarmigan species is the White-tailed Ptarmigan, an alpine specialist whose range extends from Alaska and the Yukon south, at high altitude, to New Mexico. Unlike the other two ptarmigan species, which have circumpolar distributions, White-tailed is endemic to North America and the only bird species to winter in the inhospitable alpine zone. Like its cousins, White-tailed is superbly camouflaged to blend into the tundra environment and turns totally white in winter. These birds are unbothered by human intrusion. I was once told a story by a birder searching for White-tailed Ptarmigan across an alpine hill-side in Rocky Mountain National Park. Needing to deal with his morning intake of coffee, my birding friend directed a stream of urine upon a presumed rock, which to his surprise and delight, raised itself upon two legs and crept away. He had failed to spot the bird at his feet! Ptarmigan are reluctant fliers no matter what dangers or indignities assault them but, in flight, are strong (if short-distance) fliers.

American Goshawk

Formerly called Northern Goshawk (*Accipiter gentilis*), this rough and tumble street brawler of a raptor is the largest and most formidable of the *Accipiters*. Favoring mature forest with a closed canopy and denuded understory, these shadow-colored raptors breed and winter across northern forests from Newfoundland to Alaska north to the taiga–tundra interface, and some birds have even been found nesting in mature willow thickets north of the Brooks Range. They are known for their fierce demeanor and relentless pursuit of prey—imagine if you will, a hawk the size of a Red-tailed with the agility of a Sharp-shinned and the tenacity of a Harris Hawk. If snowshoe hares have nightmares, goshawk is what they look like. And while snowshoe hare is the goshawk's primary prey, grouse, ptarmigan, and waterfowl also figure in the bird's diet.

Adult birds are slate gray above with pale gray underparts etched with a mesmerizing array of fine lines. The bird's blood-red eyes accentuate its fierce demeanor. Goshawks are mostly perch hunters, making short flights between strategic perches below the forest canopy although they also cruise-hunt forest edge and lake shores, silently gliding in on set wings. Flushed prey is overtaken in a burst of speed, and the tenacious goshawk has even been known to pursue prey on foot in woodland tangles. Once secured in a goshawk's talons, the hawk continues to "foot" its prey until it is dead, dead, dead. In northern forests in winter, there is no margin for error. One chance to secure prey may be all a bird gets in an abbreviated winter day.

While widespread across northern regions, goshawk is nowhere common, their numbers limited by the availability of prey, although the greatest threat to goshawk is not prey availability but timber harvest in the mature forests the birds depend upon. The large stick nests are often situated in the crotch of a tall tree away from roads but close to forest edge. Because of its need for privacy, the US Forest Service places American Goshawk on its "sensitive list."

While mostly a resident species, every ten years or so, goshawks engage in widespread irruptions, or "invasions," triggered by prey shortages when grouse and hare numbers simultaneously enter their cyclic lows. These irruptive flights propel mostly young birds well south of goshawk's normal winter range in the northern forests. In the West, birds can be found at higher altitudes, south even to Mexico. This larger, darker "Apache" Goshawk (*A. atricapillus apache*), is sometimes considered a candidate for separate species designation. The larger Eurasian Goshawk (*A. gentilis*) was recently separated from our North

American goshawk and is now regarded as a species distinct from birds found in North America.

Among falconers, the goshawk is a popular but temperamental bird. Only experienced falconers can fly them effectively. In Medieval times goshawk was known as "the cook's friend," meaning a dependable hunting partner. One of the few hawks bold enough to launch attacks at humans. If you stray too close to a goshawk's nest, your indiscretion will be made abundantly clear to you with close passes made by a loudly screaming adult, *yah . . . yah . . . yah . . . yah.* Attesting to the bird's fierce demeanor, the Asian conqueror Attila the Hun had the image of a goshawk emblazoned on his helmet—'nuf said?

Rough-legged Hawk

Its legs feathered to its feet, the "hare-footed hawk" (*Buteo lagopus* as translated from Greek) is admirably suited to endure harsh, cold climates. Breeding farther north than any other buteo species, this open-country hunter nests up to northern Ellesmere Island, the limit of summer's seasonal advance, and winters as far north as southern Canada (including Newfoundland). Among buteos, only the burley Red-tailed Hawk approaches Rough-legged Hawk's northern winter range. Well-adapted to open, treeless terrain, this aerial hunter can fashion perches out of thin air by hovering tirelessly over treeless terrain and studying the ground below with rodent-calibrated eyes and a swiveling head. Its small feet, too, are perfectly sized for the capture of small mammals such as lemmings, voles, mice, and shrews. Weighing a mere 2.2 pounds, the Rough-legged is able to perch on springy branches should it choose to perch-hunt for prey. Favored winter territories include marshes, pastures, wet meadows, and semidesert.

A late migrant, often waiting for snow to fly before setting out. A deep prey-concealing snowfall late in the winter may prompt Rough-leggeds to move farther south even into February, just before birds begin their spring migration, which spans

March to early May. Some have suggested that the birds lead nomadic lives in winter, but in my experience, these birds appear to settle in for the entire winter where rodent numbers are high. In the early 1800s, Alexander Wilson noted large numbers of birds wintering in the marshes south of Philadelphia, although he mistakenly considered the light- and dark-morph birds to be separate species. They are mostly solitary hunters (favored perches include utility lines, irrigation equipment, and hay bales), but with the day's hunting completed, up to fifteen birds may roost semicommunally in stands of trees. Greater concentrations (up to 200 birds) have been reported where rodent numbers are particularly high. While widely distributed across the United States and southern Canada, the Rough-legged's winter stronghold appears to be Montana and Idaho. While its population appears stable, Rough-leggeds now are only rarely seen in the marshes, but Alexander Wilson once found them to be common winter residents.

Owing to its penchant for consuming small road-killed animals, the bird is vulnerable to automobile collisions and (sadly) illegal shooting remains a persistent challenge. The bird is a mouser. You need not worry about your pheasant stock.

Purple Sandpiper

It would be an oversight beyond redemption not to give this northernmost wintering shorebird special recognition. Breeding on Arctic tundra as far north as summer is permitted to advance and where eternal winter suffers life to prevail, these chunky, fist-sized sandpipers begin their southern migration in October and November, weeks after most of their shorebird kin have fled the Arctic. Apportioning themselves in groups of ten to twenty birds along rocky coastlines from Newfoundland to the Carolinas, these birds feast mostly upon tiny mussels. South of Montauk, New York, the birds seek out man-made jetties and rock seawalls, eschewing the mudflats and sandy beaches favored by most sandpipers. A few

Purple Sandpipers do not migrate at all. Those birds breeding in southern Greenland are believed to be resident, and birds breeding in the Canadian Archipelago winter in northern Europe using Greenland and Iceland as steppingstones to Norway and Russia, where wintering birds are found north to Svalbard (78 degrees north) and the Taymyr Peninsula, Siberia (74 degrees north). (For reference, the Arctic Circle lies south of both at 66 degrees 34 minutes north.) Once established in their winter quarters, the sturdy, nimble sandpipers spend their day scrambling over ice- and seaweed-slickened rocks just above the surf but within the splash zone. This is where they seek out small mollusks (an abundant and dependable winter food resource in these northern waters), which the birds bolt down whole then pulverize in their muscular gizzards. Short legs and a low center of gravity help the birds maneuver, and if a particularly large wave does dislodge birds, they nimbly scramble to higher ground or loft into the air in a spray of dark-bodied birds to swing offshore and resettle back upon their launch point once the wave recedes. If tagged by a wave, the birds are capable swimmers but, by far, most birds can scramble to safety before being doused. At high tide, the birds may explore wadded seaweed for invertebrates, or they may cluster in tight flocks where, in more southern portions of their winter range, they may be joined by Dunlins, Sanderlings and Ruddy Turnstones.

Northern Gannet

The Sulids (boobies and gannets) are a family of large (up to 37-inch, 7-pound) fish-eating seabirds that are found mostly in tropical and subtropical regions. One notable exception is the Northern Gannet, a near albatross-sized, torpedo-shaped, white-bodied bird whose spectacular dives in pursuit of fish may reach speeds of 60 mph and carry birds fifteen feet below the surface. Flying as high as 100 feet over the water, they use their keen binocular vision to spot prey near the surface. Once prey is sighted, the feathered missiles fold their wings back as if in prayer

and plunge, bill first, into the water. Concentrating where school-
ing fish like mackerel and menhaden are concentrated close to
the surface, the sight of scores of plunge-feeding gannet arrowing
into the water is mesmerizing. An Escher painting in motion.

Eschewing both Arctic and tropical waters, these cliff-nesting
seabirds winter coastally from Nova Scotia to the Carolinas, less
commonly to Florida and Gulf waters (November to February).
Often visible from shore, feeding just beyond the breakers, strings
of migrating birds rise and fall like roller-coaster cars. After a
winter over Atlantic continental shelf waters, the adults return to
their nesting cliffs in eastern Canada (April to August). Occupy-
ing every available ledge, the bird-packed gannetries have
become tourist attractions in places like Bonaventure Island,
Quebec. But in winter, visitors to Atlantic coastal beaches can be
treated to the sight of these grand seabirds right from shore, from
Long Island to Cape Hatteras.

In winter the white-bodied adults usually outnumber
gray-bodied immature birds. Feeding concentrations often
gather where whales are driving fish to the surface, so observers
may be treated to a dual wildlife spectacle: breaching whales and
plunging gannet. The outer banks of North Carolina are a particularly good place to observe wintering gannet, but in
migration, hundreds of gannet per day may be seen in November
through December off the New Jersey coast, and Delaware Bay
supports massed numbers of "staging" northbound birds in April.

Common Poorwill

This small, eight-inch, two-ounce nightjar of open, arid, western
environments is the only bird species known to hibernate, a trait
evidently well known to the Hopi people of the American Southwest, whose name for the bird, *holchoko*, means "the sleeping one."

The "discovery" of a bird snugged into a crevice in the
Chuckwalla Mountains of southern California in 1946 shocked
the scientific community that had all but dismissed the long-held

belief that birds hibernate in winter. The inert poorwill's temperature registered 64.4°F, approximately 42 degrees below normal, and its breathing was indiscernible. Handling, shouts, and bright lights all failed to stir "the sleeping one" out of its deep torpor. Maintaining its quiescent state, the bird registered minimal weight loss over the course of the winter, and poorwills are evidently able to remain in slumber for two to three months, approximately as long as a groundhog hibernates. The Chuckwalla bird returned to the same rocky cleft in 1947, evidently unfazed by the attention it was getting. Hibernating poorwill have also been discovered in New Mexico, but poorwills breeding in the northern portions of their range appear to withdraw south in winter.

Anna's Hummingbird

Most of North America's eighteen hummingbird species withdraw into Mexico and Central America as winter enfolds the north, but not Anna's. Once restricted to extreme southern California, in the 1970s, these portly, raspberry-cowled hummers engaged in a range expansion that ferried some birds to British Columbia, where a handful choose to winter. A few birds even relocate to west-central Idaho in November through December, and banding studies have shown individual birds returning to the same inland locations on successive years.

The northward expansion has been facilitated by the planting of non-native flowers and hummingbird feeders maintained by homeowners after local breeding hummingbird species have departed. In winter, Anna's Hummingbirds also ingest insects and tree sap. As temperatures fall, the tiny birds lay down layers of fat that add to their reserves and bolster their winter hardiness. But it is the bird's ability to enter a state of torpor (semihibernation) that permits them to survive bitterly cold (subzero) nights huddled in a shrub or bush close to the food source that they will visit at first light. The birds have even been known to enter a state of energy-saving metabolic quiescence while perched on feeders.

While Anna's has earned the title of "Winter Hummingbird" (with multiple birds visiting my Californian in-laws' Central Coast feeding station in winter where nighttime temperatures often drop below freezing), several other hummingbird species, too, over-winter in the United States, among them, Ruby-throated Hummingbirds in southern Florida, Buff-bellied Hummingbirds in the Rio Grande Valley of Texas, and the Blue-throated Mountain-gem of southeastern Arizona. Snow and subfreezing temperatures can be accommodated by hummingbirds so long as they have a dependable food source. But when temperatures fall, hummingbirds may become especially defensive about sharing their nectar source. The late Sally Spofford of Portal, Arizona, once described two male Blue-throated Mountain-gems grappling in the snow despite multiple feeders at their disposal. Neither bird was injured but the food-frantic antagonists were seen even to stomp upon each other, driving their rival into the snow.

Especially in the Gulf States, hummingbird feeders maintained all winter have attracted hummingbird vagrants of assorted species that more typically winter south of the United States.

Virginia Rail

At first and even second glance, this small but widespread, mostly freshwater rail would hardly seem like a poster child for winter fortitude, yet this frail marsh bird has taught me more about bird survival at winter's rim than almost any other species. Back in my teens I used to haunt the marshes surrounding the Morristown Airport, part of an expansive marsh complex that included the Great Swamp National Wildlife Refuge, Troy Meadows, and the swamps behind my parents' home. Rich in cattails and bisected by drainage ditches cut to keep airport runways dry, the marsh hosted several wintering raptors, my principal focus on those winter expeditions. Then, one frigid winter evening while heading home, I was surprised by a Virginia Rail that strode out of the frozen marsh and began foraging along a sun-softened

bank and narrow lead of water between the ice-covered ditch and snow-covered marsh. Mostly meat eaters, the rail probed the mud and picked at the water with its forceps-like bill, finding now and again some morsel it downed with a gulp. After several minutes, its sortie completed, the rail darted back into the reeds. I took to studying the bird on subsequent evenings and it (or another bird) never failed to appear. While I never saw more than one bird at any time, the volume of tracks in the mud and snow suggested the presence of multiple rails. In winter, Virginia Rails tend to cluster, and falling temperatures concentrate them where open water is found.

Years later, living now in South Jersey, I was able to renew my study of wintering rails. Our territory on the Cumberland Christmas Bird Count encompassed a section of tidal marsh that bordered the woodland ecotone. It was our party's assigned task to "get the rail." Despite harsh winter conditions, multiple birds wintered in the reeds of the brackish marsh and could usually be coaxed into vocalizing by imitating the bird's quacking "grunt" call or by making a loud squealing call. Our responding bird's vocalizations would then incite a flurry of calls elsewhere along the salt marsh edge. Up to a dozen rails in all.

During particularly cold stretches when the entire marsh (salt and brackish) froze over, the rails would cluster in a tidal creek that ran through a pipe beneath the road. The heat from the sun-warmed asphalt coupled with the rush of water kept this narrow stretch of water ice free. Not far from the pipe was a sheltered, south-facing brackish pond that, warmed by afternoon sun, remained partially unfrozen along the northern rim even during the coldest of times. Multiple species used this life-saving resource during cold snaps, among them rails, kingfishers, Hermit Thrushes, Great Blue Herons, and Boat-tailed Grackles that feasted upon the thawed carcasses of small minnows. One bitter afternoon I even saw a Sora navigating the water's edge, my only January sighting of this tiny rail in New Jersey.

But what I learned from Virginia Rails over the years is that, for many species, winter survival is not contingent upon an abundance of life-sustaining habitat; they instead rely upon having a sustaining amount of dependable habitat when winter closes in. Not all birds are so fortunate. After a deep blanketing snowfall, I once found a weakened Virginia Rail wandering around in pine woods two hundred yards from the marsh edge. I'm not prepared to say the bird was befuddled. There were freshwater seeps in those woods that were able to sustain American Woodcocks most winters, so perhaps the rail was seeking for one of these. But my rail still faced a hard slog through sterile woodlands and deep snow. I took pity on it and elected to take it back to the marsh and release it beside one of the larger, still-flowing tidal creeks. Catching the bird was easy, and when released, it scurried up the creek without a backward glance or a by-your-leave. Laissez-faire cuts both ways.

Downy Woodpecker

It's the woodpecker next door. At seven inches, this mighty mite among woodpeckers is barely sparrow sized, yet it can winter as far north as Fairbanks, Alaska, and has the most extensive range of any North American woodpecker. Able to occupy a variety of forested habitat, it is absent only in tundra regions and the desert southwest where it is supplanted by the similar Ladder-backed Woodpecker.

Downy is less bound to mature forest habitat than many of its woodpecker kin. Wherever you feed birds in winter, this salt shaker–sized, black-and-white woodpecker is almost certainly one of your feeder regulars, particularly if you include suet in your food offerings. While year-round residents, in winter Downys may join mixed-species feeding flocks, but the moment you hoist your newly filled suet feeder into place, the bird magically appears. A biologist friend living in Fairbanks, Alaska, informs me that he has two pairs of Downys working his feeders in addition

to Hairy Woodpecker. Remaining together all winter, male and female Downys reduce direct competition for food (mostly insect larvae) by foraging in different parts of the tree, with females targeting trunks and stouter branches while males forage among outer, spindlier branches as well as gall-infested rank vegetation like goldenrod. In addition to the larvae of wood-boring insects, the Downy Woodpecker augments its winter diet with berries like poison ivy, flowering dogwood, and sumac.

Roosting in snug winter cavities in (often punky or rotting) tree trunks and thicker branches, roosting birds also readily adopt bluebird houses, whose 1- to 1½-inch-diameter holes suit their size. I have sometimes wondered whether the heat generated by the slow oxidation of rotting wood contributes to a warmer microclimate within a roosting cavity.

Being early breeders, Downy Woodpeckers are already drumming to establish their territories by February or March. In tree-impoverished areas, the birds have been known to nest in wooden fence posts; otherwise standing dead timber is the nesting substrate of choice. On cold winter nights the birds appear able to enter a state of torpor while roosting, a mechanism that helps them throttle back on energy demands to survive the prolonged darkness and subzero temperatures of far north winters. To attract Downy Woodpeckers, leave standing dead timber about your home and keep your suet feeder full. Meal worms are also relished by this species.

Brown Creeper

Among our winter birds, this tiny, gaunt, tree-hugging wisp of a bird is both surreptitious and unique. I cannot begin to recall the number of times I have wandered through winter woodlands that I pronounced devoid of life only to have my ears tickled by the high, thin call of this cryptic and clandestine mite.

Wintering across much of North America, north to Newfoundland and southeastern Alaska, the wren-like bird is almost always found in mature forest with a preference for trees with rough bark; only the bird's jerky up-hitching movements distinguish it from the bark it clings to. Short legs splayed and long, stiff, pointy-tipped tail feathers anchoring the bird in place, a creeper pries into cracks and crevices for spiders and insect larvae with its nutpick-like bill. Often craning its head sideways in the search. Built for a vertical existence, foraging birds start at the base of trees then, using a spiraling search pattern, climb with forward-set eyes focused upon the bark ahead. Running out of tree, the bird flutters like a falling leaf to the base of a nearby trunk, avoiding as much as possible long-distance flight. Upon landing, the bark-colored bird freezes, becoming one with the tree. Assured that its relocation has not attracted hunting eyes, the bird soon begins another spiraling climb.

Solitary for much of the year, in winter, creepers may join mixed-species foraging flocks, letting the eyes of flock-mates keep watch as the industrious birds keep theirs on the tree ahead. To keep warm, creepers may huddle and roost communally. Otherwise, their energetic feeding behavior serves to keep creepers warm, plus the food they find tucked in places beyond the ingenuity of other birds. Just as I so often was, observers are most often alerted to the presence of this wisp when they hear the high, quavering, vapor-thin call, *Seee*.

Winter Wren

This tiny, brown, hyperactive shrew in feathers is one of eighty-five mostly New World wren species, only two of which winter in northern regions, the other being the closely related and recently split Pacific Wren, a resident of coastal Alaska and British Columbia. The Winter Wren is the size and shape of a ping-pong ball but with a short up-cocked tail and a nutpick for a bill. Furtive, energetic, and always on the move, the tiny gremlins spend their lives on, near, even under the ground, ever searching for invertebrate prey. The only time the birds seek elevation is when breeding males position themselves in tree interiors to unleash a cascading torrent of notes whose array and volume seem impossible for a bird so tiny.

Breeding for the most part in northern coniferous forest from Newfoundland to the Yukon, in winter the birds sequester themselves across deciduous woodlands (usually wet) across eastern North America, north to the Great Lakes and southern New England. They spend their days hopping and scurrying from stumps to hollow logs to exposed root tangles in search of spiders, insects, and their larvae. Observers are often alerted to the bird's presence by its frequently given, double-noted *jip, jip* call. A solitary feeder, the notion of "flock" seems not to exist as a concept in the mind of "wren." Which is what the bird is called in Europe. Just "wren."

In the winter of 1975, I spent multiple days navigating the raised boardwalk that cuts through a patch of wet woodlands in the Great Swamp National Wildlife Refuge in northern New Jersey. It was a bitterly cold winter with a blanketing snow that lingered into March, but my regular passage seemed ever to coincide with the daily forging pattern of a single hardy Winter Wren that scurried between fallen logs, along red maple root tangles, and beneath the boardwalk itself, finding there enough unfrozen water and invertebrates to meet its high energy demands. Surprisingly and despite the richness of the habitat, the wren was often the only bird I encountered on these forays, cementing my appreciation for the hardiness of wrens. The closely related Pacific Wren is even a resident on the Aleutian Islands, where the tiny birds forage in heath and brush much like the "wren" of Europe.

Owing to their ground-foraging predilections, Winter Wrens are particularly vulnerable to roaming cats and any diminishment of old-growth forest.

American Dipper

Imagine, if you can, a plump, gray, thrush-like bird that can maneuver through fast-flowing mountain streams, not just dive and emerge like a kingfisher, but amble, totally submerged, across the stony bottom. You have just conceived of dipper, our only truly aquatic songbird and one of the planet's most specialized creatures. Tennis ball–sized and shaped, this crop-tailed denizen of western torrents can secure aquatic invertebrates (especially caddis flies) that lie beyond the ingenuity of other birds. Clad in more feathers than any comparably sized perching bird, equipped with long legs and strong feet, the bird finds traction on slippery stream beds. Dipper's short, broad wings even permit it to fly underwater and it seems especially drawn to swift-flowing rapids and cascading waterfalls. On land the nervous birds bob rapidly and habitually, giving rise to the name "dipper." One of five dipper species occurring on the planet, all of which occur in fast flowing streams, the American Dipper is found from northern Alaska to Central America but does not occur east of the Rockies.

Breeding at altitudes up to 12,000 feet, in winter some American Dippers may descend to lower altitudes while remaining in the same watershed. In northern parts of their range, the birds may also relocate to deeper rivers and lakes that remain ice free; otherwise wintering birds are found in mountainous terrain and swift-flowing watercourses year-round.

Canada Jay

There is scant need to search for Canada's national bird when you enter their territory; the plumpish, inquisitive, altogether amiable birds find you. Gliding in on short and rounded wings as silently as shadows cross the snow, the plumpish birds just suddenly appear, in your face, penetrating black eyes sizing you up for any opportunity you may present to this indicator species of boreal forest. Known by a dozen nicknames including "moose bird," "camp robber," and "whiskey jack," a derivation the name *wiskadjak* from the Cree, who consider the jay a trickster. Formerly called Gray Jay, the bird was returned to its original name of Canada Jay by the American Ornithological Society in 2018. The genus name *Perisoreus* means to "heap up," a clear reference to the bird's penchant for caching food items, which it glues to the undersides of bark using its sticky saliva. Food caching is essential for birds to survive the privations of the long northern winter and may explain why these birds choose to nest in February, the coldest month of the year. An early start to the breeding season gives juvenile jays more time to cache the food they will need to survive their first winter.

Omnivorous and inquisitive, the birds will explore open tents, pilfer bacon frying in the pan, and rally to the sound of a gunshot, anticipating that a meal of fresh, warm viscera is in the offing. Meat hanging in hunting camps belongs to the jay that finds it, but the birds also eat wasps, mice, and berries.

The penchant for this jay to nest in the dead of winter, while not unique, is at odds with most songbird species. Ravens and Mourning Doves, too, are early nesters, but most boreal forest breeders wait until spring to raise young.

Golden-crowned Kinglet

Weighing in at a mere six grams, less than some North American hummingbird species, it seems impossible that this tiny, acrobatic pixie can secure enough food to see it through the winter, but across much of North America, north even to southern Alaska and Newfoundland where the birds are resident, the hyperactive Golden-crowned Kinglet finds a way. It dances on the tips of branches in search of tiny invertebrates too small to interest most bird species. Favored prey includes small insects, springtails, spiders, mites, and their eggs. The birds seem ever on the move, probing among short-needled conifers or hovering at the tips of branches, plucking food beyond the nimble reach of most other gleaning species, including chickadees. Kinglets are enabled by their small size, which allows them to penetrate pine needle clusters, and their light weight, which permits them to perch even on springy pine needles. Ever in motion, even when stationary the birds flick their wings nervously.

In winter, kinglets form groups of two to six birds that then team up with mixed-species foraging flocks dominated by chickadees. Observers are often alerted to the bird's presence by its hyper-high-pitched contact call *ti, ti*—a call whose frequency (8,000 Hz) lies above the range of many human adults. Able to survive in temperatures well below zero, the birds have been known to roost in the protective confines of squirrel nests and are believed to roost communally, in clusters, as their Old World counterpart the Goldcrest is known to do. Some authorities consider the Golden-Crowned Kinglet and Goldcrest to be, in fact, conspecific.

Despite their hardiness, prolonged cold and snow may result in high local mortality among kinglets. In winter, birds occupy pine forest and hardwood bottomland and generally prefer mature trees with a particular partiality shown for sweet gum. In mixed-species flocks dominated by Carolina Chickadees (another hover-gleaner), the birds may seek to avoid competition by foraging slightly away from or behind the main flock. While the Golden-crowned is a boreal forest breeder whose range extends from Newfoundland to Alaska, maturing pine plantings south of its historic range resulted in a twentieth-century expansion of its breeding range to even Maryland, New Jersey, and Illinois.

Foraging consumes most of the bird's daily wintertime budget, and while showing a modest long-term population decline, birds in the east seem to be increasing. In winter the birds sometimes feed on suet.

Hermit Thrush

North America's six spot-breasted thrushes are all celebrated for their ethereal flute-like songs, but can be difficult to separate, especially since the forest gremlins seem invariably to be facing away, concealing their distinguishing underparts. However, in winter, the challenge of thrush identification across North America is greatly simplified. From December through March, the only spot-breasted thrush found north of the tropics is the Hermit Thrush, our winter thrush. Hardy and widespread, this common denizen of the forest understory winters across much of the southern United States and Mexico, north to British Columbia and southern New England—switching over from a summer diet of invertebrates to mostly fruit. In southern New Jersey, the bird is one of the most common winter denizens of the forest understory, spending the colder months ensconced in berry-rich greenbrier tangles, especially those mantling a trickle of water or a snow-melting seep.

When heavy snow blankets the ground, the birds may relocate to roadsides bracketed by trees where they forage through the leaf litter exposed by snowplows. There, they find protein-rich invertebrates to supplement their winter diet. Flying at car-grill heights, relocating from roosting to feeding areas at dawn and dusk, hurried commuters may not see the darting form in the headlights in time to react, and another member of the forest community becomes a statistic, one of the millions of birds killed annually along US roadways. In the west, Hermit Thrushes winter in lush gardens with water sources, making them particularly vulnerable to predation by roaming cats. Nevertheless, Christmas Bird Count totals these past decades have shown a stable population.

American Robin

Roger Tory Peterson's one-line description of (Eastern) Robin in his famous 1934 *A Field Guide to Birds* reads: "The one bird everybody knows." But do you?

Everyone's picture of this brick-breasted thrush is a bird with a wiggling earthworm in its yellow bill striding across short-clipped lawns or a bird singing its heart out on the branch of a blossoming apple tree. Iconic, yes; accurate, perhaps not so much. A robin perched on a snow-draped American holly with a berry in its bill would be more apt because over much of North

America, for much of the year, this "harbinger of spring" is a tough winter survivalist.

True, across most of Canada and Alaska American Robin *is* absent in winter, but the stout thrush breeds in mature willow thickets north of the Arctic Circle and winters across almost all of the lower forty-eight states and south into Mexico. Indeed in the American southwest, South Texas and most of Florida, winter is the only time you'll find "America's favorite bird" (as christened by naturalist Dallas Lore Sharp). The birds you see in the summer, foraging for invertebrates like earthworms, are not necessarily the same birds found in your region in winter, whose dietary focus has by then shifted from worms to fruits that they find in forest interiors, orchards, and along highway center divides, not in your yard. In winter, the sense that the "robins have left us" relates mostly to a shift in foraging tactics and diet, not to the species' physical relocation. Robins do not eat bird seed so are unlikely to come to your feeder. But robins are compulsive bathers and will use bird baths year-round. A heating element placed in the bath will ensure open water in colder regions, or you can replace water daily.

Chances are, there are hundreds of robins wintering in your vicinity, roosting in stands of trees or marsh reed. These winter residents fan out to explore food-rich areas at first light and mass to roosting areas before sunset. Winter aggregations of 50 to 500 robins are common and roosts of 250,000 birds are known. I estimate the number of robins wintering in southern New Jersey to range between 200,000 and 500,000 birds, which roost in the extensive white cedar stands flanking the region's tidal wetlands. By day the birds forage on the bounty of holly berries that festoon the forest understory; greenbrier berries, grapes, poison ivy, and juniper berries are also prized winter food items.

On warm January days, the birds may even break into song.

The switch over from meat to fruit is a gradual process. Where I used to deer hunt in Hunterdon County, New Jersey, 50 to 200 American Robins would forage daily in the snow-dusted leaves of the forest floor in front of my tree stand, in search of earthworm egg cases hidden between the leaf litter and ground. At dusk the birds would retreat into the adjacent stand of planted white pines. This pre-roost feeding frenzy occurred at least into the first two weeks of December.

By then deep snow fully covered the ground and the birds had entirely switched over to feeding on berries, but on sunny days a few robins could always be counted upon to forage along the bare leaves and earth exposed on the sun-warmed south side of the planted pines.

In spring, robins migrate north on the heels of the 37.4°F isotherm. When ground temperatures rise to 40°F, hibernating earthworms will return to your yard and with them your resident robins, heralding the return of spring with open beaks bubbling with song. *Cheerily, cheer up cheer up cheerily, cheer.* Often beginning at first light, across North America no seasonal fanfare is more welcome or apt than the caroling song that is the source of one of the robin's nicknames, "the wake robin." The moniker is often attributed to celebrated naturalist and essayist John Burroughs, whose 1871 book about the natural history of birds bears this storied title.

Yellow-rumped Warbler

All but a handful of North America's fifty-four warbler species seek out the warmth of the tropics in winter, and then there is the Yellow-rumped Warbler, which not only winters widely across much of the United States but even north to Nova Scotia, British Columbia, and southern Ontario. Outwardly, there is nothing to suggest that this tiny boreal forest breeder is made of sterner stuff than its kin, but thrive across the winter landscapes it does, by the tens of thousands in places like the bayberry thickets of coastal Carolinas, the leaf-holding oaks of California's Central Coast, and the vast southeastern forests where Yellow-rumps team up with Eastern Bluebirds and Brown-headed Nuthatches among the loblolly pines. Key to the Yellow-rump's winter hardiness is its versatility. From gleaning to flycatching to foraging on the ground to plucking prey from pond surfaces to engulfing berries, Yellow-rumped Warbler does it all. Able to metabolize the waxy covering of bayberry and consume poison ivy berries, the birds can survive in temperatures that immobilize insects. Moving through bayberry thickets like a hungry cloud, filling the air with their flat *chep* notes, and flashing their signature yellow rumps, scores of birds might be flushed out by an observer from a particularly berry-laden thicket on North Carolina's Outer Banks. Cedar berries, too, are relished by the bird. In California both the Myrtle and Audubon's subspecies of Yellow-rumped thrive in grape vineyards and enliven mature oaks with their nonstop gleaning, which on warm days turn, once again, to catching insects on the wing.

Migrating later in the autumn than most warblers and able to withstand temperatures well below freezing, the hardy Yellow-rumped most certainly wins the crown for pluck and resourcefulness.

McKay's Bunting

This exceedingly rare and geographically restricted Arctic denizen spends its life hugging the shores of the Bering Sea, breeding almost exclusively on the isolated Bering Sea's Hall and St. Matthew Islands, wintering coastally in western Alaska between the Seward Peninsula and the Alaska Peninsula. Males look similar to Snow Bunting but overall whiter in their breeding plumage with just touches of black in the wingtips and tail. In winter groups of up to twenty-five McKay's join flocks of Snow Bunting to forage for weed seeds on coastal beaches.

My personal fascination with the bird began with a childhood perusal of the National Geographic Society's two-volume *Birds of North America*, published in 1965. Presuming I would never have the opportunity to see such a rare and isolated bird, I was nevertheless captivated by the idea of a pure white songbird. But by exceedingly good fortune, in 1987 my wife, Linda, and I were invited on an adventure cruise, which made a stop at remote St. Matthew Island. Separating ourselves from the group, we pursued a male McKay's carrying insects to its subterranean nest, settled in a rocky cleft on a denuded outcropping overlooking our landing beach. There we spent the next two hours lying with chins pressed to hands enchanted by the comings and goings of the industrious little bird, its bill brimming with insects. A male skylark displaying over the beach below was an unexpected bonus.

Presumed not to winter on the islands, the buntings spend mid-December to mid-March on the Alaskan mainland, staging on the shores of Hooper Bay due east of their breeding grounds from April 30 to May 20. In a 2018 survey, the world population was believed not to have exceeded 19,481 adults, down from 31,560 individuals counted in a similar survey in 2003. Their very small population and geographic isolation makes McKay's Bunting extremely vulnerable to any environmental disruptions but especially predation by any introduced mammalian predator such as foxes or rats—the bane of isolated island bird populations everywhere. The island homes of the birds fall under the jurisdiction of the Alaska Maritime National Wildlife Refuge, which controls human access. The buntings were named by ornithologist Robert Ridgeway in 1884 based upon two specimens collected in winter by army officer C. L. McKay near his remote post on Nushagak, Alaska. The breeding grounds of the bird were not discovered until 1887, when naturalist Charles H. Townsend collected both adults and juveniles on Hall Island in early September.

Red-winged Blackbird

There are many bird species that might claim to be the harbingers of spring. American Robin is, perhaps, the popular favorite but in far northern regions the return of American Crow antedates the robin's arrival by several weeks. In Florida, the first returning Purple Martins reach North American airspace as early as early February. But for my money, spring's most iconic herald is the widespread Red-winged Blackbird. Gathered in large flocks across the United States and parts of Southern Canada north to Alaska and Nova Scotia—all winter the birds lead circumspect lives, foraging among the reeds and open ground, creeping like penitents, trying not to draw the attention of the bird-eating hawks that blackbird flocks attract. Then one morning with warm southerly winds pushing cumulus clouds across the sky, everything changes. Game on!

The male blackbirds, who have kept their incendiary red epaulets discretely furled all winter, are suddenly moved to take some high perch and throw caution to the wind. Drawing their heads back, letting crimson epaulets blaze, the birds utter the gurgled incantation that picks winter's lock: *Tur-a-ling! Tur-a-ling!*

It's a glove thrown right into the face of winter that asserts loud and clear, "I live. Wanna make something of it?"

Where most creatures seek only anonymity, on this, the first day of spring, male Red-winged Blackbirds throw caution to the wind in the name of moving their genes forward and assert themselves.

To a female Red-winged Blackbird, *Tur-a-ling* translates as "Hi ladies. Mr. Right, here."

To rival males, it says: "Listen up guys, the biggest, burliest Red-winged Blackbird in the place claims this corner of the planet. Beat it."

Of course, such a boast cannot go untested. Up and down the marsh edge, from every elevated perch, other male Red-wingeds hitch themselves aloft and mimic the performance of their rival, each blazing epaulet a spark that kindles a flame, flames that ignite a firewall to beat back winter's stalled advance.

"Sorry, not sorry," the wall of red shields signals. This is as far as you go, Winter. You've had your season and now, you've shot your bolt. From this day forward, every sunrise dawns closer to summer. We won. You lost.

Tur-a-ling!

Black-capped Chickadee

American Robin may be, in the eyes of North America's residents, our most recognizable bird, but everybody's favorite bird must be the pert, winsome chickadee. Nimble, acrobatic, vocal, and confiding, if you live in North America, there is almost certainly a chickadee species (or close relative) near you. And most likely your neighborhood tit is the amiable and widespread Black-capped Chickadee, a resident of northern forests from Newfoundland to western Alaska.

South and east of Black-capped's range, the species is replaced by the slightly smaller but equally winsome and nearly identical Carolina Chickadee. In the west, Black-capped is supplanted by the rakish Mountain Chickadee and colorful Chestnut-backed Chickadee. All chickadee species are social, amiable, and readily attracted to backyard feeding stations where they favor black-oil sunflower seeds. Indeed, if you hang a feeder in your yard, the very first visitor will almost certainly be one of these bold, inquisitive birds. The spirited sprite may even land on feeder perches before the feeder stops swinging.

While chickadees are not dependent upon offerings of seed, in colder regions feeding stations do almost double their winter survival rate by providing reliable, high-energy food on demand when energy-taxed birds need it most, dawn and dusk. When temperatures fall to 10°F, chickadees that are able to supplement their diet with offered seed survive at near twice the rate of feeder-deprived birds. Weighing in at a mere half ounce (12 grams), the birds are challenged to maintain their body heat, given their surface area to mass ratio. For years scientists marveled at the tiny bird's ability to survive nighttime temperatures that in parts of the Black-capped Chickadee's range may plunge to −50°F. Part of the answer is strategy, part physiology. Northern Black-capped Chickadee may be 25 percent larger than their southern kin—a heat conservation adaptation—and chickadees have a layer of insulating down feathers that are denser than most comparably sized songbirds. Also key to Black-capped's success is its ability to accommodate a variety of habitats and exploit various food resources, including boreal forest, deciduous forest, marsh reed, and broken suburban habitats.

Preparation for winter begins long before cold weather sets in, with birds caching hundreds if not thousands of seeds for later retrieval. The birds have remarkable memories, aided by a seasonally enlarged hypothalamus. So cerebrally endowed are chickadees that the birds can store and remember the hiding places of up to half a million seeds and insect larvae. Consider that the next time you go searching for your car keys.

Spending most of their daily time budget foraging for food, the birds consume up to 60 percent of their body weight before going to roost, and over the course of a day add 8 to 20 percent of their weight in body fat, which will be depleted by morning. Roosting cavities rank second only to food security in terms of the chickadee's winter survival strategy. In central Alaska, birds may spend up to eighteen hours roosting and have as little as six hours of foraging time per day. Roost cavities are excavated in the trunks or branches of standing dead trees, whose punky cores are easily worked by the birds. Birch trees are especially favored and cavities just large enough to snugly accommodate a single bird are typical. Other bird species roost communally, but chickadees like to sleep alone. On especially cold nights, the birds are able to enter a state of energy-saving torpor, which reduces body temperatures by 12 to 15°F and their metabolic demand by 25 to 32 percent.

Despite this metabolic throttle back, by dawn the birds will have exhausted their fat reserves and will require about twenty minutes of warm up to get their body temperatures up to the requisite 108°F operational range before setting out to feed. The birds accomplish this metabolic warmup by shivering. Then it's off to join up with the winter flock (which average six to eight individuals but sometimes up to fourteen flock members) to begin another abbreviated day foraging across a winter territory that may encompass forty acres, using a route that hits all prime feeding spots, including and probably starting with the feeders in your yard. Despite their energy-deprived state, it is not a free-for-all at feeders. Chickadees maintain a strict feeding hierarchy (a pecking order) with senior flock members (the alpha pair) feeding first, while other members watch for danger and wait their turn. During the day the birds remain warm by exercise and sunning themselves. Over much of its range, the chickadee's primary nemesis is the Sharp-shinned Hawk, a robin-sized woodland predator with rapid acceleration and quicksilver reflexes.

But chickadees are far from defenseless. Not only do flocks maintain constant vigilance, but neighboring flocks communicate a predator's approach by adding *dee* notes to the *chickadee-dee-dee . . .* sequence. A four-to-five *dee* sequence is chickadees being conversational. Additional *dee* notes signal danger. The more notes, the greater the threat. So where an eleven *dee* sequence may signal a prowling cat a twenty *dee* tirade indicates a very serious threat, like a perched Northern Pygmy-Owl or inbound Sharp-shinned Hawk. Heads up! A single *zee* call means "Freeze, danger imminent." Later, a conversational *chickadee* call uttered by a senior flock member signals, "All clear."

The chickadee social structure is highly sophisticated. Beginning in March alpha pairs split off from winter flocks and establish breeding territories in the core of their winter territory. After the breeding season, the alpha pair begin mustering their winter flock, which consists of juveniles raised by neighboring pairs plus free-floating adults. This chickadee core then constitutes the base of mixed-species foraging flocks that wander through winter woodlands.

Another far northern chickadee species well adapted to survive in the harsh winter environment is the plumpish and overall browner Boreal Chickadee. If you live outside the boreal forest, you may never see this species. Ranging farther north than Black-capped Chickadee (even to 50–60 degrees north latitude), the northern limit of the Boreal's range coincides with the limit of the white spruce forest that sustains them. North of the boreal forest, you are entering the rarified domain of North America's rarest and hardiest chickadee, the Gray-headed Chickadee (aka Siberian Tit), which lives in stunted spruce forest along the south slope of the Brook's Range and breeds in mature willow thickets along major Arctic river courses. Utilizing the entire array of chickadee survival tricks, the densely feathered Gray-headeds regularly encounter temperatures of −30 to −50°F in January and February, with daylight limited to one to two hours of twilight per day. Temperatures as low as −70°F have been recorded in this region, making this northernmost mountain range one of the coldest places on Earth.

Retreating again farther south, where the ranges of Black-capped and Boreal Chickadees overlap, Boreals choose to forage in the denser interior of spruce forests and are generally near the tops of trees. Less vocal and social than Black-capped, Boreal's *chickadee* phrase is slower and more slurred.

Boreal Chickadees do, occasionally, engage in irruptive flights into the lower forty-eight states, where individual birds may team up with foraging bands of Black-capped Chickadees.

To see a Gray-headed Chickadee, you are best advised to take an organized raft trip down some Arctic river in summer when the birds are nesting. Go armed: willow thickets in the Arctic are frequented by grizzly bears, who may have cubs and don't like surprises.

Sparrows—Meditations in Brown

Upon reviewing this manuscript, I discovered that I had inadvertently given sparrows short shrift. An indefensible oversight insofar as, arguably, no bird family is more representative of winter across North America than this dapper, New World bird group. Breeding all across northern North America, many species, like the American Tree Sparrow and Dark-eyed Junco, herald from Arctic and boreal regions. In winter, the fifty or so sparrow species apportion themselves widely across the United States, southern Canada, and northern Mexico. Cryptically plumaged and ground hugging, the birds in this group might easily be overlooked except for their frequently given vocalizations and tendency to flock. Songs range from sonorous to loud and assertive; call notes range from frail and lisping to sharp and explosive to brittle as ice.

Not as colorful as goldfinches nor as acrobatic as chickadees, sparrows are nonetheless a vibrant component of the winter birdscape, with many wintering species arriving just as the last of autumn's leaves shed their branches and departing as the first buds of spring begin to swell. In winter, many sparrows are flocking birds, roosting and foraging in loose aggregations, enlivening the bleak winter days with their songs and calls. Specialized for life on or near the ground, most sparrows prefer to feed beneath rather than upon commercial bird feeders. Nevertheless, flocking species like White-throated and White-crowned Sparrows may constitute the most abundant birds at your feeder and are mostly punctual about their arrival and departure dates.

Roger Tory Peterson once observed of the two chestnut-capped sparrows that frequented his Connecticut yard, American Tree Sparrow and Chipping Sparrow, that he could identify the birds simply by looking at the calendar. The blush-colored Tree Sparrow, an Arctic breeder, is a winter resident, found in Connecticut from November through February; the similar but more crisply attired Chipping Sparrow is a summer breeder, departing by October. Any chestnut-capped sparrow seen in Connecticut on the 4th of July is invariably a Chipping Sparrow, all Tree Sparrows by this mid-summer date are sequestered in willow thickets and stunted spruce across Arctic and subarctic regions where their sharp, clear-whistled song melds with the canary-like chatter of Common Redpoll and the somnambulant chant of White-crowned Sparrow to harmonize into the Arctic's summer symphony.

In winter, in the lower forty-eight, American Tree Sparrows are drawn to weedy patches with a shrub component. The pert, rakishly capped Chipping Sparrows, which winter across the southern United States and Mexico, prefer more denuded habitat, often foraging on suburban lawns and the sparse understory of southern pine forest where in winter they may join wintering flocks of bluebirds, Pine Warblers, and Brown-headed Nuthatches. The birds gravitate toward conifers, favoring the touch of pine needles underfoot.

Across much of North America, winter's onset is heralded by the arrival of sparrows and ends with their departure. It is not without basis that across much of North America, juncos are widely known as the "Snow Bird." Yes, despite their slate gray upper parts, a junco is a sparrow, and one whose winter range encompasses most of North America, from southern Canada to northern Mexico. Indeed, this dapper sparrow might easily serve as the poster child for winter. Arriving in October, departing in April, the Snow Bird lives up to its name. But before departing in the spring, the ground-hugging birds post notice of the changing season by taking high perches on sunny mornings and filling the crisp winter air with a musical trill that flies in the face of winter and is richer and more melodious than the brittle trill of the Chipping Sparrow, which will be arriving any day.

But juncos are just one of the sparrow species that enliven winter days. In my youth, the common winter sparrow was the lively American Tree Sparrow, with dozens of birds infesting the weedy edge of my parents' suburban yard and massing to my bird feeding tray (actually, our patio table) where the birds devoured about half my weekly allowance of twenty-five cents. What I got for my money was a winter landscape enlivened by the birds' high, brittle call notes. I have long believed that if ice had a sound, it would recall the cheery *tchee, tchee, tchee* of the American Tree Sparrow.

Residing most winters, now, in South Jersey, our common winter sparrow is the portly White-throated Sparrow, birds of woodland edge that scratch the snow beneath our feeders for spilled seed, and every evening fill the air with their explosive call notes until the chorus swells to a crescendo as birds retire for the night into the protective confines of our Olympic-class forsythia. The bird's evening fanfare always recalls to me the final scene in the old TV series *The Waltons*, where family members chorused their good nights to each other as shadows settled: "Good night, Grandpa . . . good night, Jim Bob . . . Good night, Elizabeth . . . good night, John Boy . . . good night . . . good night . . ." or as expressed in sparrow speak: *chink . . . chink . . . chink . . . chink . . . chink . . . .* Taps for another winter day as our troop of sparrows takes roll call. On any sunny winter day, male White-throated Sparrows may break into their trademark song, a whistled lament to *Old Sam Peabody, Peabody, Peabody.* Which to my ears, sounds more akin to *Oh, say, say, see, see, see.*

In California, in winter my in-laws' ranch is infested with White-crowned Sparrows, a trim, athletic cousin of White-throated Sparrows. The birds' sonorous and somnambulant dirge is integral to the winter soundscape. Swarming beneath the feeders by day, in the evening the birds retreat into the safety of rose bushes and planted pines where their whistled song signals the end of another day. Scattered among their feeding ranks are plumpish Golden-crowned Sparrows. While White-crowneds are widespread in winter—found all across much of the United States, far western Canada, and northern Mexico—Golden-crowneds are more restricted, relegated to brushy habitat along the West Coast from British Columbia to Baja.

While most sparrows are cryptically garbed, clad in conservative browns, buffs, and grays, some are dapper and distinctive. Among the dapper few is the harlequin-faced Lark Sparrow, a prairie specialist that in winter drapes itself over utility lines then drops to weedy roadsides below. More restricted but equally distinctive is the robust Harris's Sparrow. A burley, black-bibbed, brush-hugging sparrow that breeds in the tundra–taiga interface and winters in America's heartland, where it sings throughout the winter.

Not to be forgotten are the towhees, which despite their name and sometimes colorful plumage are just elongated woodland sparrows. Both the western Spotted Towhee and Eastern Towhee brandish rust-colored sides, and while sometimes difficult to see in their brushy enclaves, towhees are easily detected by their double scratch sequence as they forage among the fallen leaves. After blanketing snows, towhees and other woodland sparrows often gather along roadsides where snowplows have exposed fallen leaves. Drive with caution.

Other sparrow species are specialized to forage in grasslands and deserts. One of the most common and widespread of these is Savannah Sparrow, a short-tailed, streak-breasted denizen of sparsely vegetated open areas, able to eke out a living where other sparrows might starve. One of the bird's many subspecies is the pallid beach-loving Ipswich Sparrow, which breeds on Sable Island and ekes out a living on sparsely vegetated upper beach between New England and Georgia. Given the bird's limited and exacting habitat requirements, the population of roughly 6,000 adult Ipswich Sparrows is closely monitored.

Conversely, the somewhat similar but longer-tailed and mostly solitary Song Sparrow has one the most extensive ranges of any sparrow, being found coast to coast and border to border. With twenty-four subspecies, Song Sparrows boast a variety of plumages. Most races are rufous-tinged above and darkly streaked below with a signature dark spot over the heart, but southwestern birds are overall paler with pinkish streaks. The large, lead-colored Aleutian race of Song Sparrow is barely recognizable as a sparrow. But no matter where you live in North America there is likely a Song Sparrow near you to enliven the winter landscape, including in vacant urban lots.

House Sparrow

Many readers are likely surprised that the spunky ubiquitous House Sparrow was not included in my sparrow profile. The reason is simple. The suggestion of its name notwithstanding, the House Sparrow is not a New World sparrow. It is a weaver finch, related to the weaver finches of Africa. An immigrant to our shores, the House Sparrow is now found wherever we are, a gregarious, streetwise urchin as much at home in the streets of Brooklyn, where this Old World native was first introduced in 1851, to the Plaza Square in Salt Lake City where it was introduced in the 1870s.

At home in feed lots in Colorado, desert rest areas along I-40, suburban yards, and even inside airport terminals. No accounting of winter birds would be complete without reference to this spunky little bird whose affinity for bird feeders is surpassed, perhaps, only by chickadees; indeed, in urban habitats, House Sparrows may be the only songbird species attending your feeder. Less finicky than chickadees, almost any food offering is grist for this granivore mill.

Vilified by bluebird and Purple Martin landlords who decry the sparrow's nest box–usurping predilections, the bird is nevertheless an American success story. An immigrant introduced to these shores that flourishes wherever people do.

Breeding for the most part in cavities, including the nooks and crannies of buildings, the House Sparrow's northern expansion was understandably stymied by the limits of the boreal forest and the resulting absence of tree cavities. At present, House Sparrow's northernmost outpost is Churchill, Manitoba, a grain-exporting center for Canada's wheat growers. Here at the very northern limit of the boreal forest, the ever-resourceful immigrant species has taken up residence in the port city's grain silos, which afford the birds protection from the Arctic winter as well as a near-unlimited food supply. Otherwise, in winter, House Sparrows gather into flocks of thirty or so birds and forage for food during the day, roosting at night in trees, shrubs, and protecting structures. Once sustained by the undigested oats they found in horse droppings on city streets, when automobiles replaced draft horses, this resourceful weaver finch shifted its dietary focus from undigested grain to the processed grain bound up in discarded donuts and pizza rinds. America's profligate eating habits have been a boon to this hardy, street-tough little bird.

AN IMBOLC BIG DAY

I t is one thing to assert that North America hosts a wealth of wintering birds. It is another to demonstrate it. Here I set out to affirm this assertion, first by tabulating those birds that have survived the winter around our winter retreat in Paso Robles, California, and then, across North America by enlisting the help of friends and colleagues who accepted my challenge to conduct two-hour-long surveys of the birds near their homes on Groundhog Day—an Imbolc Big Day intended to take the pulse of the season.

Accordingly, on February 2, the absolute end of winter, I and others set out to conduct tallies of the birds wintering within walk-ing distance of our respective abodes. The date was chosen to coincide with the Druid Imbolc celebration, conducted February 1 to 2, the astronomical mid-point between the Winter Solstice and the Spring Equinox. Also known as Groundhog Day, on this date (February 2) winter is turning the corner. The sun creeping ever higher in the sky will reach its zenith on June 21, but on February 2, our star is just reaching the mid-point in its annual ascent into spring. On February 2 those birds that elected to winter outside North America have yet to return, so all birds counted in North America on February 2 are winter residents. And winter survivors.

Having an entire day to dedicate to the endeavor, I chose to conduct my Imbolc Big Day from dawn to dusk recording all those birds found here on the Central Coast of California where my wife, Linda, and I spend our winter. Way back on December 21 (the first day of winter), the ranks of birds, here and elsewhere, were bloated by first-year birds, juveniles who had yet to face down a northern winter. But now, in February, the ranks of wintering birds have been whittled down to the hardy survivors—birds that have earned the right to call themselves winter residents and move their genes forward in spring.

My survey area consisted of my in-law's five-acre Central Coast property and the adjacent vineyards and pastures within easy walking distance of their ranch house door. Observations began at 6:00 a.m. outside the guestroom window and ended at 6:00 p.m. near the hot tub/pool.

In deference to their time constraints, the other participants of the Big Day were invited to limit their bird counts to two hours—thus, a sample of birds in their areas, not a census. While not comprehensive, their efforts would take the pulse of season and, in sum, provide a broad picture of wintering birds across North America.

Every bird tallied this day was a hardy survivor. A winter warrior whose ingenuity and fortitude allowed them to face down winter and stand on the verge of spring.

Paso Robles, California, February 2, 6:00 a.m.

My personal Big Day effort dawns upon a Central Coast trapped in a classic winter-rain pattern, with Pacific moisture drawn up into an "atmospheric river" that slammed the state February 1, dropping locally three inches of rain with more to follow. Southern California receives its annual allotment of rain in winter. Snow in the higher elevations.

Happily for me, February 2 dawns with the Central Coast poised between storm systems. While it rained most of the night, precipitation ended before dawn with temperatures falling to 42°F (cool by Central Coast standards). The sudden temperature drop coupled with the rain have precipitated a thick layer of tule fog with visibility falling to about fifty feet—not the most auspicious of conditions for bird "watching" but better than pouring rain. And while tule fog tends to be persistent, rather than postpone my count I decide to gamble that the fog will burn off by late morning. Fog or no, the birds are here and will be all day (just as they have been here all winter). In addition, no matter what the weather, my colleagues across North America can be counted upon to turn out to count the birds in their regions. Furthermore, the forecast for California called for worse weather to follow—multiple days of torrential rain. This morning's fog, while challenging, is not damning. Given the day-long scope of my effort, my hopes are pinned upon a productive afternoon and (especially) a fruitful evening as birds settle in to roost. As events unfold, and as predicted, California is indeed inundated by multiple days of rain (ten inches in some places) and widespread flooding. But February 2 (count day), while foggy early, sees clearing skies and productive birding most of the day.

Lingering in bed, listening to the drip, drip, drip of water off the roof, I begin my tally by studying the oak outside the guest-room window and letting my ears do most of the work as I enjoyed the warmth trapped beneath the quilt and a mug of coffee. Punctually, at 6:00 a.m., our local troop of California Quail announce the new day with a fanfare of flock calls, *wa-ha-her* . . . *wa-ha-her* . . . *wa-ha-her*. Often phonetically rendered *Chicago* . . . .

Then, single file, hugging the fence line, the lead birds line out toward my mother-in-law's garden where they will spend the day. Stepping quickly, top nots bobbing . . . the strikingly patterned, grapefruit-sized birds file past, eyes alert for the hunting hawks that know the morning routine as well as they.

Meanwhile, a continent away in Western Pennsylvania, hundreds are gathered on Gobbler's Knob to witness the emergence (extraction?) of the celebrated weather-predicting groundhog, Punxsutawney Phil, to determine whether his Royal Prognostication will see his shadow and predict another four weeks of winter, or his shadowless emergence will foretell an early spring.

Having no groundhogs west of the American Great Plains and uncertain about the jurisdictional limits of an eastern rodent, anyway, I am relying upon our West Coast wintering birds to pass judgement on the season's progress. My second bird of the day is the Northern Mockingbird, whose scratchy call notes serve to pass judgement upon the human preoccupation with ascribing meaning to natural events.

Wa-ha-her . . . wa-ha-her . . . wa-ha-her . . . crow the California Quails—a dozen birds in all.

Back in November, the wintering flock had numbered sixty-five individuals but over the course of the winter, the hunting hawks had cut into their ranks. Today's tally will suggest by how much. As if on cue, the lusty screaming of a Red-shouldered Hawk reaches my ears. *Kee-yer, kee-yer, kee-yer. . . .* While these forest hawks are not an especially dire threat to quail, John James Audubon's somewhat sensationalized painting of a juvenile Red-shouldered does show the bird, talons splayed, flailing among a wildly flushing flock of the Northern Bobwhite. Across much of New England, the lusty cries of Red-shouldered Hawk are the fanfare of spring. The "singing hawk," in the words of ornithologist Leon Augustus Hausman.

But our resident Red-shouldered's focus this morning is on procreation, not quail. In February California Red-shouldered Hawks rekindle their pair bond. The male bird is celebrating the onset of the breeding season and the coming winter's end by crying lustily. Of far greater threat to the quail are our resident Cooper's Hawks, who are on this day of my bird count unaccountably absent. On most mornings, the juvenile Cooper at least could be counted upon to perch in the oak outside the window, studying the quail as they pass. But this morning, no hawks cast their shadows upon the flock. Also unaccountably absent are the hordes of Yellow-rumped Warbler that have, for most of the winter, swarmed the leafless branches of the (magic) oak in search of insect eggs and larvae. Even the local Bushtits are absent—the buzzing flock feeding elsewhere. I surmise that the oak twigs, picked over all winter by hungry birds, are finally devoid of food. The warblers and tits year have moved on to greener pastures. On this chilly 42°F morning, it is certain they are feeding somewhere else and will all day. My task is to find them.

The next birds to make the roster are a tidy swarm of White-crowned Sparrows followed by a trio of Mourning Doves, who settle onto the pasture already greening up from the winter rains, as are the northern hills, still shrouded in morning fog. Tucked up against the Coastal Range, Paso Robles sits on the western edge of the verdant San Joaquin Valley. Once an oak-dotted savannah, with an annual rainfall pushing a desert-like ten inches per year, the local grasslands have, over the years, largely given way to vine-yards, and pecan and pistachio orchards. Adjacent to my in-law's ranch are several thousand acres of grapes, whose fecund rows support wintering Yellow-rumps, hordes of House Finch, and flocks of blackbirds and doves, along with a smattering of Western Meadowlarks. The birds are drawn to the sugar-rich bounty of unharvested grapes and, for much of the year, the drip lines sup-plying the vines with the fossil water drawn from the underlying aquifer. North America's agricultural lands are a boon for many wintering bird species, and in addition to producing some of the planet's finest wines, the Paso Robles wine region supports mil-lions of wintering birds—some like the Yellow-rumped Warbler are northern breeders, others like the House Finch, year-round residents. While now found across all of North America and likely the most common bird at your feeder, the House Finch was his-torically a bird of xeric habitats in the southwestern United States and Mexico. Illegally sold in the east under the trade name "Hol-lywood Finch," escaped birds came to establish in the northeast in the 1940s, beginning a westward expansion which culminated in the eastern interlopers joining up with their native kin by the 1990s. Our California birds are almost certainly natives.

The morning parade of quail over, and seeing patches of blue through the fog, I quit the covers and relocate to the front porch of the ranch house to see what birds are swarming the feeders on this chilly morning. House Finch and House Sparrow dominate. Both species acclimate well to human environs, including urban and suburban areas. Our local troop of twenty House Sparrows spends most of the day perched in the ornamental shrubs below the kitchen window, where they have both protection and the feeders in view. Two Collared Doves foraging beneath the feeders fly off at my approach. Predictably on such a cool morning, the humming-bird feeder is busy, hosting at least two male Anna's Hummingbirds, down from the six birds that were here back in December. Then:

Chlah . . . chlah . . . chlah . . . chlah . . . chlah.

Our local troop of California Scrub-Jays are jump-starting their day by moving through the oaks in search of acorns. They are dedicated pack animals, and rarely will you encounter a single Scrub-Jay. Moving through the branches like a troop of feathered squirrels, the birds vault across open areas with the coordination of a SWAT team. Two at a time—one bird leading, one bird trailing just off the lead bird's wing—the same formation used by combat fighter pilots (and for the same defensive reason).

But I've located some of my missing Yellow-rumped Warblers. Still clad in their mid-winter gray plumage (nearly devoid of yellow highlights), a dozen birds are dancing among the outer branches of the elms, accompanied by a pair of Oak Titmice, our local tit species. No chickadees here east of Paso, except for the occasional Chestnut-backed Chickadee forced down from the hills by drought.

The thistle socks outside the kitchen window are festooned with Lesser Goldfinches, and beneath the bird bath is an adult Golden-crowned Sparrow feeding among the many White-crowneds. Curiously, I have not seen a Golden-crowned all winter. An auspicious day for the bird to show up. Thanks for making my list, fella. Have a safe trip back to Alaska. You're about two months out and then it's on to a busy breeding season at the edge of Alaska's coastal spruce forests—not that you need me to tell you how to go about the business of being a bird. Your presence, here, on this late date assures me you've got the right stuff.

Woodpeckers the world over are late risers but by midmorning they are now out in force, with the clown-like Acorn Woodpeckers dominating. A dozen boldly patterned birds are cavorting among the trees whose ring-like configuration traces the foundation of the original property owner's pioneer homestead. The Acorn Wood-pecker's raucous *rah . . . rah . . . rah . . .* call approaches the level of a din, and their treetop antics recall a Keystone Cops routine as the birds vie for favored perches and access to the suet. Looking closely at the tree trunks, you'll see the mosaic of holes drilled into the bark by these industrious woodpeckers, each excavation sized to fit the acorn the birds hoard to see them through the winter. Look closely and you'll note that many of the holes now are empty, their con-tents exhumed and consumed over the course of the winter. The dearth of cached acorns accounts for the bird's focused attention upon the suet block, already half consumed by starlings.

Our other resident woodpecker is Nuttall's Woodpecker, a near California endemic closely related to Downy and Ladder-backed Woodpeckers. But I cannot linger to see them; patches of sunlight are already burning through the fog. A good sign. With no high overcast to block the sun's warming rays, the fog will melt, salvaging my Big Day. It's time to beat feet; I still have lots of ground to cover and birds to tally.

Strolling down the driveway to Buena Vista Drive, I flush two California Towhees and the resident pair of California Thrashers, who even at this early date are already taking an interest in the rose bush below the kitchen window into which they snug their twig nest. The apple trees in the adjacent garden, still burdened by fruit, have attracted a swarm of White-crowned Sparrows, which are feasting upon the pulpy sweetness of fist-sized Pink Ladies, still crisp and juicy despite the late date. Elsewhere across North America, last autumn's apples are dried and shriveled, but here in the climate-controlled conditions of a California winter, the fruit remains produce-counter fresh—a boon to wintering birds, for-aging mule deer, and appetite-driven writers. Among the sparrows are two Cedar Waxwings and a single American Robin. These two fruit-eating species habitually team up in winter, and apple orchards are a favorite draw everywhere in North America.

Now, on the main road, with horse pasture on the right and vineyards on the left, I line out in the direction of the Le Vigne tasting room. While much too early for a glass of wine (even by California standards), the tasting room's outside porch is a fine place to scan the skies, which are beginning to clear and drawing soaring birds aloft. Maybe I'll work in a glass of Central Coast Chardonnay on the return trip. But oh . . .

Deer . . . deer . . . deer . . . deer . . . deer . . .

My musings are suddenly interrupted by Killdeer erupting from the rain-soaked pasture, drawn there, I presume, by earthworms driven to the surface by the rain. It's too early in the season for migration and migrating Killdeer seldom gather in flocks of more than a dozen birds, anyway. No, these are wintering birds concentrated, this morning, by the opportune food resource. The flock circles and lands, offering ample time for study, but I can find no Mountain Plover in their ranks. Thirty years ago, when grazing cattle were common in this region, the "Prairie Ghost," as the bird is known, was an annual winter resident here. Now these endemic North American shorebirds are rare, their winter habitat usurped by grapes, whose rows now support House Finch, blackbirds, and Yellow-rumped Warblers.

The adult Red-tailed Hawk perched atop a roadside oak ignores the smaller birds, its eyes and talons calibrated for the many ground squirrels that infest the pasture. California's Central Coast is a Red-tailed Hawk stronghold, and resident birds are already courting. With thermals beginning to perk, some birds are aloft, engaged in swooping, leg-drooping courtship displays. Also taking advantage of the free ride after a day of rain are a dozen ravens and a Golden Eagle. Yes, birds flying over a count area on a Big Day count too.

Having missed seeing any Western Meadowlark, I continue past Le Vigne and spot a yellow-breasted bird perched atop a

pistachio tree. But my presumed meadowlark proves to be a Cassin's Kingbird. A bonus species. At the northern limit of its winter range, the flycatcher leads a precarious existence, its survival contingent upon the severity of the winter. Indeed, the two Cassin's that made my New Year's Day yard list a month before appear not to have survived the January cold spell. I haven't seen them for days. The bird in the pistachio is an unexpected bonus, and my scrutiny leads to another discovery: a distant flock of 2,000 or so blackbirds and starlings foraging among the grapes. Come evening the birds will file by the ranch en route to their roosts down by the creek, but it is more convenient to get a tally now. Study brings me to conclude the flock was three-fourths blackbirds (mostly Red-wingeds), balanced by starlings. Approaching mid-afternoon, I begin retracing my steps, stopping off at Le Vigne just as clouds begin to roll from the west. Enjoying the view south over the vineyards from the patio, I am treated to the cavorting of White-crowned Sparrows in the shrubs and the Yellow-rumped Warblers dancing through the planted olive trees mantling the patio. Hoping to get a tally of Turkey Vultures before the clouds shut down thermal activity, I count ten. Then, after sparing with a rival neighbor, a soaring Red-tailed Hawk sets its wings and descends toward its mate perched next to a newly constructed nest. Their mating, while perfunctory, was evidently successful, and the pair remain together near the nest.

Amanda, one of the tasting room's attentive servers, notes my presence and, recognizing me as a regular, asks whether I'd like "anything"? Hesitating only slightly, I agree that a glass of Chardonnay would be a fine addition to the day. I sip slowly, and the glass lasts long enough for the utility lines to begin filling up with Red-winged Blackbirds staging to go to roost and a dozen Brewer's Blackbirds strutting across the parking lot. Predictably, two other patrons confide that the gathering reminds them of (the movie) *The Birds*. Boy, I hate that stupid, baseless film. As unflattering to birds as *March of the Penguins* was not.

Declining a refill and realizing the afternoon is moving on, I start back to the ranch, hoping to pick up some of the species missed earlier.

Back on the porch, scanning the tops of oaks for Phainopepla (and the missing Nuttall's Woodpeckers), I am treated to the antics of the resident White-breasted Nuthatches. The hummingbird feeder is rarely vacant as Anna's Hummingbirds tank up for the night. There appear to be three hummingbirds now. As the afternoon wears on, the Siberian elms begin filling up with blackbirds, whose conjoined clamor sounds like a cross between a calliope and a pet shop at feeding time.

Suddenly, *silence*. Not a reduction in volume. Silence, as every bird vies not to be the one present. Evidently a hawk has made an appearance. Probably one of the resident Cooper's Hawks, who could hardly let a gathering of birds this size go untested. But insofar as the cacophony resumes immediately, I am brought to wonder whether the hawk was not, instead, a flyby Prairie Falcon or Merlin. But during the silence, I realize that the White-crowned Sparrows have already begun their evening Vespers. It's time to relocate to the pool area, where nightly roosting birds pack into the surrounding roses for the night. The Phainopepla will have to go down as a "no-show."

Away from the blackbird din, the sonorous dirge of White-crowned Sparrows fills the air as evening shadows reach across the yard and meld into darkness. Right at dusk, a Black Phoebe arrows in to snap insects off the surface of the pool. Breeding in the shed above the hot tub, the resourceful flycatcher evidently survived the winter's several hard frosts. As the sun dips below the Coast Range, an avalanche of White-crowned Sparrows vaults the house en route to their rose-bed fortress. Festooning themselves like Christmas tree ornaments across the outer branches, lifting their bills to the sky, I note that few juvenile birds are in the ranks now. Most of the birds I scan with binoculars appear to be adults, and many are in full song.

*Deeee, dee-da-lee, dee, dee, der . . . dee-da-lee dee, dee, der . . .
dee-da-da-lee . . .*

Come April, the White-crowned Sparrow's signature song will
fill the Arctic—reverberating from willow thicket to willow
thicket. But in winter, the birds and the song belong to the Central
Coast. And now as the day's last sunlight gilds the tops of clouds,
their sonorous lament marks the close of my Imbolc Big Day. At
precisely 5:00 p.m., according to the pool-side clock, five California Quail rocket into the roses, the typical evening pattern. As they
burrow into the latticework of thorny branches, sputtering the
flock sounds of quail tucking themselves in for the night, I am
dismayed by the paucity of birds. In December, my evening quail-
roost counts sometimes topped sixty-five birds. Had so many quail
failed to survive the winter? It is nearly dark, and I stand to leave,
eager to see whether any hummingbirds remain at the feeder.

But halfway across the yard, the rumble of wings brings my
head around in time to see a tight-packed volley of forty quail
melting into the rose bush. Evidently, the flock's roost schedule
has changed—a pattern shift, imposed, perhaps, by hunting
Cooper's Hawks. Now, the bulk of the birds are arriving after
Cooper's Hawk hunting hours and the birds are flying en masse,
a technique used to confuse hunting eyes. The quail count stands
at forty plus five—a total of forty-five, then. It appears that only
twenty or so birds have failed to survive the winter—a nominal
loss that ensures both the survival of the quail and the hawks that
hunt them. As I continue toward the house with the last of the
White-crowned Sparrows still singing their soulful dirge. I realize that I've gone the day without hearing our local Great
Horned Owls who, in February, vocalize dawn and dusk.

Pausing, studying the branches of the elms, I am delighted to
see the shadow of an owl weave through the branches and perch
above the drive. Bringing binoculars to bear upon the head
turned my way, I see no ear tufts. A Barn Owl, I am forced to
conclude, drawn to the rodents that will soon be searching for

spilled seed beneath the feeders. The owl's focus on rodents is greatly appreciated by local landowners. Artificial Barn Owl nest boxes are a common fixture along the edge of vineyards and here, at least, the Barn Owl population remains healthy. My Imbolc Big Day over, I retreat into the house to count up my earnings and catch the Weather Channel's recap of news from Gobbler's Knob.

Full-Day Count Total (38 species; 2,502 individuals): Mallard (1), California Quail (45), Rock Pigeon (13), Eurasian Collared-Dove (12), Mourning Dove (20), Annas's Hummingbird (3), Killdeer (60), Turkey Vulture (14), Red-shouldered Hawk (2), Red-tailed Hawk (8), Golden Eagle (1), Barn Owl (1), Acorn Woodpecker (8), Nuttall's Woodpecker (2), Northern Flicker (3), American Kestrel (3), Black Phoebe (3), Couch's Kingbird (1), California Scrub-Jay (6) Common Raven (10), Oak Titmouse (3), White-breasted Nuthatch (4), Bewick's Wren (1), Western Bluebird (4), American Robin (1), California Thrasher (2), Northern Mockingbird (3), European Starling (500), Cedar Waxwing (3), House Sparrow (20), House Finch (40), Lesser Goldfinch (5), California Towhee (4), White-crowned Sparrow (120), Golden-crowned Sparrow (1), Red-winged Blackbird (1,500), Brewer's Blackbird (15), Yellow-rumped Warbler (60)
The Great Horned Owls didn't begin calling until after midnight, so did not make the count.

Regional Imbolc Big Day Mini-Counts

The following two-hour bird tallies were conducted by cooperators, within walking distance of their respective homes on February 2.

Fairbanks, Alaska
Observer: Ted Swem, 8:56–11:00 a.m.

Habitat: Regenerating spruce forest with some birch, on a rural suburban street, plus bird feeders. Observations conducted of feeders from inside the house. Observer estimates an actual number of up to 75 chickadees and 75 redpolls being served by his feeder in a given day.

Weather: Overcast with no wind, −30°F.

Total Count (5 species, 46 individuals): Downy Woodpecker (2), Hairy Woodpecker (2), Common Raven (2), Black-capped Chickadee (20), Common Redpoll (20)

Author's musings: Ted's feeders appear to lie at the intersection of two or more chickadee winter-flock territories. But even 20 birds at the close of an Alaskan winter attest to the winter hardiness of this species. No penguin has ever survived a winter in the Alaskan interior.

Paso Robles, California

Observer: Pete Dunne, 6:56–9:15 a.m.

Habitat: Five-acre ranch with mature oaks, Siberian elms, rose garden, apple orchard, vineyards, pistachio orchard, and also a 10-acre pasture. Multiple bird feeders.

Weather: Heavy rain overnight, ending before dawn. No wind. Dawn temperature 42°F. Dense fog with a visibility of 50 feet, clearing by 10:00 a.m. Broken overcast temperature in the 50s. No shadow until 10:00 a.m.

Total Count (26 species, 217 individuals): California Quail (5), Northern Mockingbird (3), White-crowned Sparrow (30), Mourning Dove (18), California Scrub-Jay (1), Northern Flicker (1), Red-tailed Hawk (1), Red-shouldered Hawk (1), Eurasian Collared-Dove (2), Anna's Hummingbird (2), House Sparrow (20), House Finch (18), European Starling (16), Yellow-rumped Warbler (12), Acorn Woodpecker (8), Western Bluebird (2), Black Phoebe (1), Lesser Goldfinch (2), Oak Titmouse (3), American Robin (1), Cedar Waxwing (2), White-breasted

Nuthatch (2), California Towhee (2), California Thrasher (2), Golden-crowned Sparrow (1), Killdeer (60)

Deschutes County, Oregon, Deschutes River Woods
Observer: Marina Richie

Habitat: Front yard with feeders and heated bird bath, and adjacent forest—ponderosa pine with manzanita, bittersweet, and currant berry understory.

Weather: Mostly cloudy, 36°F, intermittent and light precipitation. No shadow.

Total Count (9 species, 58 individuals): Cooper's Hawk (1), Northern Flicker (2), Mountain Chickadee (6), Dark-eyed Junco (12), Pygmy Nuthatch (14), House Finch (7), Common Raven (2), Lesser Goldfinch (13), Steller's Jay (1)

Author's musing: I am surprised by the solitary nature of this Steller's Jay—jays are a bird group I associate with flocking. Not that I question the accuracy of the tally, and I am compelled to note that I, too, tallied but a single jay on my two-hour survey.

Lafayette, Colorado, Greenlee Wildlife Preserve
Observer: Ted Floyd

Habitat: Small suburban park east of Boulder, featuring ponds, meadow, woodlands, and a picnic area.

Weather: Cloudy, cool, humid, no wind. 40°F. No shadow.

Total Count (41 species; 1,356 individuals): Snow Goose (1), Cackling Goose (562), Canada Goose (30), Northern Shoveler (17), Gadwall (8), American Wigeon (34), Mallard (70), Green-winged Teal (81), Ring-necked Duck (14), Common Goldeneye (7), Hooded Merganser (13), Common Merganser (4), Turkey (1), Rock Pigeon (20), Mourning Dove (3), Ring-billed Gull (5), Great Blue Heron (2), Bald Eagle (3), Red-tailed Hawk (2), Downy Woodpecker (1), Northern Flicker (23), American

Kestrel (1), Blue Jay (10), Black-billed Magpie (14), American Crow (10), Common Raven (17), corvid sp. (1), Black-capped Chickadee (18), Bushtit (30), White-breasted Nuthatch (4), Starling (130), American Robin (45), House Sparrow (70), House Finch (30), American Goldfinch (3), Dark-eyed Junco (7), White-crowned Sparrow (14), Song Sparrow (3), Spotted Towhee (4), Red-winged Blackbird (43), sparrow sp. (1)

Author's musings: The volume of birds is a tribute not only to the bird riches of the area but the acumen of the observer, one of North America's most celebrated field birders. This count alone should dispel any notion that bird riches do not abound in winter in the continent's interior.

Ash Canyon, Huachuca Mountains, Arizona

Observers: Tom Wood and Tim Blount

Habitat: Rural neighborhood with oak woodlands, and multiple feeders and water sources.

Weather: Partially cloudy, intermittent drizzle. 40–45°F. Observer saw no shadow.

Total Count (23 species, 94 individuals): Montezuma Quail (3), Inca Dove (6), Mourning Dove (6), Rivoli's Hummingbird (1), Anna's Hummingbird (2), Broad-billed Hummingbird (1), Cooper's Hawk (1), Red-tailed Hawk (1), Acorn Woodpecker (3), Gila Woodpecker (3), Ladder-backed Woodpecker (3), Arizona Woodpecker (1), Northern Flicker (1), Say's Phoebe (1), Mexican Jay (30), Bridled Titmouse (3), Verdin (4), Bushtit (12), Ruby-crowned Kinglet (2), White-breasted Nuthatch (1), Bewick's Wren (2), Cactus Wren (1), Curve-billed Thrasher (6)

Author's Musings: I guess this should dispel any questions about hummingbirds wintering in the US. Three species!

Alpena, Michigan

Observer: Matthew Derr

Habitat: Along Thunder Bay River, with hydroelectric plant maintaining open water.

Weather: Clear skies, no precipitation. 37°F. Observer saw his shadow.

Total Count (14 species, 181 individuals): Mallard (26), Redhead (2), Common Goldeneye (4), Common Merganser (9), Red-breasted Merganser (38), Ring-billed Gull (9), Herring Gull (1), Rock Pigeon (46), Common Raven (4), Black-capped Chickadee (26), Tufted Titmouse (1), Northern Cardinal (3), American Goldfinch (5), Purple Finch (7)

Author's Musings: The impressive waterfowl tally is most certainly a tribute to the presence of the hydroelectric facility and the open water it generates.

Granville, Ohio

Observer: Peggy Wang, 8:45–10:55 a.m.

Habitat: 350-acre Denison University Biological Reserve—mostly beech forest, with successional agricultural land with springs and flowing stream.

Weather: Cloud cover 80 percent, winds light and northerly. 43°F. Observer saw her shadow.

Total Count (24 species, 113 individuals): Canada Goose (1), Black Vulture (2), Turkey Vulture (4), Red-shouldered Hawk (1), Red-bellied Woodpecker (7), Downy Woodpecker (1), Northern Flicker (7), Blue Jay (6), American Crow (6), Carolina Chickadee (6), Tufted Titmouse (4), Golden-crowned Kinglets (2), White-breasted Nuthatch (4), Carolina Wren (8), Eastern Bluebird (9), Hermit Thrush (3), American Robin (8), House Finch (2), American Tree Sparrow (1), White-throated Sparrow (14), Song Sparrow (2), Eastern Towhee (4), Rusty Blackbird (1), Northern Cardinal (10)

Author's Musings: I am particularly intrigued by the sighting of Rusty Blackbird, a once-common wintering species of swampy woodlands that has been in serious decline for half a century.

Egg Harbor Township, New Jersey

Observer: Donald P. Freiday

Habitat: Pine Barrens suburban habitat (chiefly pitch pine, oak, and sweet gum), plus feeders and multiple clearings.

Weather: Cloudy with intermittent showers. 46°F. Winds northwest 5–15 mph. No shadow.

Total Count (23 species, 141 individuals): Canada Goose (25), Mourning Dove (18), Cooper's Hawk (1), Red-breasted Nuthatch (2), Downy Woodpecker (2), Hairy Woodpecker (1), Northern Flicker (1), Carolina Chickadee (12), Tufted Titmouse (6), Golden-crowned Kinglet (2), White-breasted Nuthatch (2), Carolina Wren (5), Eastern Bluebird (4), Hermit Thrush (2), American Robin (20), House Finch (8), American Goldfinch (8), Chipping Sparrow (3), Fox Sparrow (2), White-throated Sparrow (10), Song Sparrow (1), Yellow-rumped Warbler (4), Northern Cardinal (2)

Author's Musings. The observer's tally flies in the face of the commonly held belief that the New Jersey Pine Barrens are devoid of birds.

Cape May Court House, New Jersey

Observers: Dale Rosselet and Kevin Karlson, 1:45–4:00 p.m.

Habitat: Their yard, bracketed by marsh and woodlands, with feeders; plus a walk to Delaware Bay beaches across salt marsh.

Weather: Overcast skies. 42°F. Wind north-northwest, 5–7 mph. No shadow.

Total Count (33 species, 5,833 individuals): Red-winged Blackbird (45), House Sparrow (10), Downy Woodpecker (2), Yellow-bellied Sapsucker (1), Mourning Dove (65), Carolina Chickadee (8), Tufted Titmouse (6), American Robin (45), Dark-eyed Junco (35), White-throated Sparrow (200), Song Sparrow (3), Swamp Sparrow (1), Blue Jay (12), Grackle (5,000), Northern Mockingbird (2), Carolina Wren (2), European Starling (87), Hermit Thrush (2),

Brown Thrasher (1), Northern Harrier (2), Cooper's
Hawk (2), Red-shouldered Hawk (1), Ring-necked Pheas-
ant (1) American Black Duck (6), Bonaparte's Gull (40),
Herring Gull (20), Great Black-backed Gull (5), Great
Egret (1), Double-crested Cormorant (1), Hooded Mergan-
ser (1), Clapper Rail (1), Sanderling (25), Dunlin (200)

Author's Musings: The presence of such a large flock of wintering grackles, while not unprecedented, does make me
wonder whether this is a daily foraging pattern or perhaps a
late-season relocation of birds.

Author's Summary

In reviewing the totals, I am struck mostly by the absence of
surprises, except for Tom Wood's hummingbird list. The species
enumerated and totals posted paint the picture of a continent
poised for spring, with bird numbers appropriate for a late-winter
census. However, a few things are worth comment. Ted Swem's
Fairbanks, Alaska, chickadee total of twenty individuals and
observation that up to seventy-five may be attending his feeders
attests to not only the hardiness of this tiny bird but (perhaps) the
convergence of multiple foraging flocks at his feeders. Marina
Richie's single Steller's Jay in Bend, Oregon, strikes me as curi-
ous, but this was a census not a population survey. Donald
Freiday's three Chipping Sparrows straddles the limits of the
bird's winter range limit. Dale Rosselet and Kevin Karlson's Great
Egret likewise represents a species at the northern limit of its
winter range. Migrating egrets will not arrive in Cape May until
late March. All in all, a splendid effort on the part of a talented
cadre of field birders. Maybe we should consider making this a
Groundhog's Day tradition? Birds are more widespread than
groundhogs and cast shadows, too.

And just as a reminder, these totals reflect bird numbers following a winter's attrition. These are the survivors.

EPILOGUE
DRIVING A WEDGE
INTO WINTER

For all its hardships, winter is not forever. Birds with the right stuff have been facing down winter's hardships since the Laurentide Ice Sheet left town. The prize for having survived winter is the right to move your genes forward in the spring. Come March, waxing sunlight drives winter north, thawing ice-locked bodies of waters.

Many years ago, as I huddled beside just such a lead of winter-darkened water as fat snowflakes swirled in confusion, my vigil was rewarded by the sudden appearance of a wedge of five diving ducks that arrowed past on whistling wings and disappeared behind the curtain of snow. How the migrating birds spied the open water I'll never know.

Holding my breath, listening past the hiss of snow striking water, I was rewarded by the sound of ducks breasting the surface. Minutes passed as I projected a beam of youthful longing into the maelstrom and, magically, the birds separated themselves from the storm. Five shadow-colored birds swimming toward my huddled form. Three drakes and two hens that approached so close I could see the namesake copper ring on the necks of the males, my first Ring-necked Ducks. Wintering as far north as they find open water, these hardy diving ducks are among the first migrants to head north in spring, driving a wedge into winter with their determined passage. Among waterfowl it is males who follow the females to their natal area. And the birds were nearly home, only one or two days out, according to the range map in my field guide.

Unaware of my presence, the birds dove for dragonfly nymphs and exploded into flight only when, toes stinging and knees stiff with cold, I stood, eager to get home and consult my field guide to determine what special gift of the season had just been granted me.

Five hardy diving ducks that through pluck and luck had lived to see the spring and were in the process of bearing their genes north to pass their hardiness and fortune on to the next generation. The encounter was momentary. I knew the birds would not linger. Spring is not forever, either.

Spring advances north at a rate of thirteen miles per day (a near glacial pace), but this advance is not infinite; ultimately spring, with summer right on her heels, runs headlong into the intractable likes of the permanent Arctic ice cap. Three to four meters thick in places, this barrier to living things extends across the northernmost islands of the Canadian Arctic Archipelago and northern Greenland. Just south of this glacial barrier, summer and winter wage a two-month battle for supremacy, with the tide of battle swinging day to day. At 60 to 90 degrees latitude north, it can snow any month of the year.

On June 21, 2001, the first day of summer, my wife, Linda, and I were poised at the edge of the seasonal ice cap, whose retreat had brought the edge of winter's ice to the mouth of Eclipse Sound, the eastern portal to the Northwest Passage, which even on this first day of summer was still firmly encased in ice. South of us was Baffin Bay, east of us perhaps a mile of open ocean, the limit of spring's advance. North of us, Bylot Island was still mostly encased in snow. Ten feet before us, jammed up along the edge of the sea ice were thousands of King Eider, Long-tailed Ducks, and Red-breasted Mergansers, all

EPILOGUE

driven into a hormonally fueled frenzy of frustration because
their drive to breed had been stymied by a late spring. Until
inland tundra lakes and ponds thawed, the birds could do little
but hoot, quack, bray, hiss, and yodel their frustration. The conjoined sounds were enough to topple the walls of Jericho but
not enough to move the intractable winter ice. Or, as our Inuit
guide sagely proclaimed, "Ice is boss." In the far north you
respect this wisdom, or you die. The age lines creasing his
mahogany-colored face bore witness to his stockpile of wisdom,
and the crow's feet branching from both eyes attested to his
many hours spent squinting into a horizon-hugging sun searching for danger. Indeed, the residents of nearby Pond Inlet were
at the time of our visit mourning the loss of one of the village
elders who had gone out on the ice to hunt seals but never
returned. His gas cans, cached for the return trip, were found, but
the hunter and his snow machine were not.

Beyond the press of ducks and the open water, a polar bear
and her two cubs were prowling, their interest piqued by our
colorful cluster of tents. To the south, the tusks of Narwal flashed
in the morning sunlight, attesting to yet another stretch of water
opening between us and Baffin Island. But the ducks whose
destination lay somewhere to the north were stymied, all the
miles they had navigated thwarted by winter's unrelenting grip.
And now with the Solstice upon us, and the Earth beginning her
six-month slide into another winter the urgency of the birds was
understandable. But the year's breeding season was not canceled,
only delayed. In the days of twenty-four-hour sunlight ahead
there would be enough warmth to thaw the tundra ponds. In a
week, maybe less, the ice we stood upon would be open water.
Even, now, on the wind-scoured hilltops of nearby Bylot Island,
newly returned American Golden-Plovers were laying eggs, and
the Arctic air sizzled with the courtship flight of White-rumped
Sandpipers just back from their hemisphere-vaulting migration
to and from southern South America. Down in the valleys, water

157

gurgled beneath sun-rotted winter ice. On a wind-swept ridge bordering the inlet, brant were grazing on last summer's grass, their passage, too, stymied by the frozen landscape.

No, summer was not canceled. It was simply delayed. A late thaw is one of the risks all high-arctic breeders must sometimes accommodate no matter what their winter survival strategy. But this temporal bottleneck at the migratory finish line had drawn an entire spring migration down to a press of waterfowl.

Two to eight duck-lengths deep.

At the front of the pack, the breasts of the birds were pressed against winter. Summer was lapping at their tail feathers and their concupiscent drive was urging them onward. I've studied spring migration my entire life but always thought of it as a broad front, and protracted affair. But here in the Arctic where seasons meet, spring was compressed to a narrow ribbon of life snugged up against a stubborn wall of ice. Just beyond the ducks, out over the open water, Arctic Terns were wheeling and diving for smelt. Their 22,000-mile journey to and from Antarctic ocean waters was over. The planet's migratory champions were home and on a gravel beach. A scant five miles away females were already staking out their little corners of heaven, awaiting their mate's impending return and his offering of fish, a food transfer that would cement the pair bond, and another breeding season would begin. Sealed with a fish.

Meanwhile, out on the thawing tundra, Rock Ptarmigan were molting into their summer plumage and celebrating winter's end by feasting upon those willow buds newly emerged from their shroud of snow. The croaking of males echoed across every valley—a fitting fanfare for this hardiest of Arctic birds' latest victory over winter.

ACKNOWLEDGMENTS

N o act of writing is singular. Behind every writer is a support force whose insights and assistance facilitates each and every writing. Accordingly, and with unbridled gratitude, I wish to thank the thousands of scientists whose insights and field observations figure in these pages. In particular I want to single out my friend and retired US Fish and Wildlife biologist Ted Swem, whose many years of service out of Fairbanks, Alaska, gave him unparalleled opportunities to study the birds of the Arctic.

Also in need of recognition is Don Freiday, biologist, hunting partner, and friend, who has been a party to many of the bird encounters recounted in this book, as were mentors Floyd P. Wolfarth and Jack Padalino, who accompanied me on many a winter foray along the banks of the Delaware River in search of wintering eagles and waterfowl.

Tom Gilmore, my boss of thirty years, graciously afforded me time to travel on assorted writing projects, from when encounters with birds figure in this book. And by extension, this gratitude is accorded my colleagues at Cape May Bird Observatory, who had to shoulder extra duties while I was away from my desk. While many online resources were consulted during the writing of this book, the contribution of the Cornell Lab's *All About Birds* (allaboutbirds.org) website cannot be gainsaid. Many of the facts and behavioral facets tucked into this manuscript have their origin there.

No book of mine does not involve wife, Linda, my incomparable partner of forty years, who genially puts her life on hold while I pursue my passion for birds, writing from the Maine coast to the American Prairies, to California, to Attu Island and the Bering Sea, and the Canadian Arctic where Bob Dittrick and Heimo Korth shared with me an extended adventure on the Alaskan taiga and winter's edge. It was here that we three came upon the snow-white hare pinned to the Earth by the smoke-colored bird of prey. You know, guys, the cloaking snow that fell that night might just have saved that hare's life but also deprived the goshawk of a meal.

Gratitude is extended to those friends and colleagues who lent their talents to my Imbolc Big Day, most notably: Ted Swem, Marina Richie, Ted Floyd, Tom Wood, Matthew Derr, Peggy Wang, Don Freiday, Kevin Karlson, and Dale Rosselet.

For writers, the encounters accrued during a human life account for little unless committed to paper. Driven by this truism, my writer's retreat in Maine commands center stage, as does, by extension, the generosity of my friend and benefactor Beth Van Vleck. If this name sounds familiar, it is because you ran across it 50,000 words ago in the book's dedication. Thank you, Beth, for the extended use of your lovely home on the shores of Penobscot Bay, and for your friendship. As I write this line, I hear the morning stirrings of my incomparable mother-in-law, Ann Ellis, whose five-acre ranch in California's grape country has afforded me near unlimited insights into the winter lives of Central Coast birds, most notably the California Quail that roost in her rose bush, Mer, and who are, at the moment of this writing, filing past the guestroom window in route to a busy day of survival . . . or not.

PETE DUNNE
February 11, 2024

APPENDIX
WHERE THE BIRDS ARE

This section strives to divide North American breeding birds by their winter apportionment across the hemisphere, dividing them into four broad categories (or tiers).

These delineations are intended only to impart some measure of order to winter-bird distribution, and while based upon the known distribution, the placement of species into categories is to some degree capricious, insofar as the winter range and migratory strategy of several species give them standing in more than one category. For example, Short-billed Dowitcher, a plump shorebird species, has a winter distribution that extends from the southern United States to northern South America but is here designated an intermediate migrant (tier 3) because the destination defined by this category is where most members of this species appear to winter.

Wintering Peregrine Falcons, too, could be relegated to several different categories, with some birds wintering in the Pribilof Islands (tier 1), others in metropolitan areas in North America (tier 2), and still others in the Caribbean and Antilles (tiers 2 and 3), as well as South America (tier 4). So Peregrine Falcon might qualify as a northern resident or a short-distance, intermediate, or long-distance migrant, but focusing as I chose to do upon the bird's celebrated powers of flight (as evidenced by their scientific name *Falco peregrinus*), in my categorization I elected to celebrate the bird's capacity for long-distance flight, thus designating it a tier 4, long-distant migrant.

In the case of overlapping categories, I typically selected in favor of the species' strong suit—it's more celebrated (or extreme) wintering strategy.

Thus, if a species is advantaged to withstand the hardships of the northern winter even if a portion of the population does not remain within the Arctic region, the species is still classified a northern resident (tier 1). Semantics aside, what this method of categorization clearly demonstrates is that, by far, most North American breeding birds also winter on the North American continent (Mexico being part of North America). This determination may surprise many, including many serious students of birds, and does not in any way diminish the critical role that the forests of Central and South America play in the life cycle of tens of millions of North American–breeding birds, many of which herald from the great boreal forests of North America, the planet's greatest bird hatchery. The categories are:

Tier 1: Northern Residents (48 species). Species that breed and winter in extreme northern regions.

Tier 2: Short-Distance Migrants (487 species). Short-distance migrants and more southerly resident species that breed in the United States and Canada, and winter for the most part in the United States and northern-to-central Mexico.

Tier 3: Intermediate Migrants (69 species). North American breeders that spend the north's winter primarily in Central America and northern South America.

Tier 4: Long-Distance Migrants (67 species). Species that travel great distances to winter mostly in the Southern Hemisphere.

Tier 1: Northern Residents

The hardiest of the hardy, including those species that, in winter, remain in the Arctic region (like Rock Ptarmigan) and northern waters (like Thick-billed Murre), plus boreal-resident species (like American Goshawk and Canada Jay). These species include:

Emperor Goose	Snowy Owl
Steller's Eider	Northern Hawk-Owl
King Eider	Great Gray Owl
Common Eider	Boreal Owl
Spruce Grouse	American Three-toed
Willow Ptarmigan	Woodpecker
Rock Ptarmigan	Black-backed Woodpecker
White-tailed Ptarmigan	Downy Woodpecker
Sharp-tailed Grouse	Hairy Woodpecker
Thick-billed Murre	Gyrfalcon
Least Auklet	Canada Jay
Whiskered Auklet	Common Raven
Crested Auklet	Black-capped Chickadee
Atlantic Puffin	Boreal Chickadee
Pigeon Guillemot	Gray-headed Chickadee
Black Guillemot	Red-breasted Nuthatch
Rock Sandpiper	Bohemian Waxwing
Glaucous Gull	Pine Grosbeak
Ivory Gull	Common Redpoll
Ross's Gull	Hoary Redpoll
Red-legged Kittiwake	Red Crossbill
Yellow-billed Loon	White-winged Crossbill
Red-faced Cormorant	McKay's Bunting
Pelagic Cormorant	Snow Bunting
American Goshawk	

Tier 2: Short-Distance Migrants

Encompassing US permanent residents and short distance migrants, these include species wintering north up to southern

Canada, and south to central Mexico and the Bahamas (many of these are short-distance migrants). R = resident species; M = migrant.

Black-bellied Whistling
  Duck (R)
Fulvous Whistling-Duck (R)
Snow Goose (M)
Ross's Goose (M)
Greater White-fronted
  Goose (M)
Brant (M)
Cackling Goose (M)
Canada Goose (M)
Mute Swan (R)
Trumpeter Swan (R/M)
Tundra Swan (M)
Muscovy Duck (R)
Wood Duck (M)
Blue-winged Teal (M)
Cinnamon Teal (M)
Northern Shoveler (M)
Gadwall (M)
American Wigeon (M)
Mallard (R/M)
American Black Duck (M)
Mottled Duck (R)
Northern Pintail (M)
Green-winged Teal (M)
Canvasback (M)
Redhead (M)
Ring-necked Duck (M)
Greater Scaup (M)
Lesser Scaup (M)

Harlequin Duck (M)
Surf Scoter (M)
White-winged Scoter (M)
Black Scoter (M)
Long-tailed Duck (M)
Bufflehead (M)
Common Goldeneye (M)
Barrow's Goldeneye (M)
Hooded Merganser (M)
Common Merganser (M)
Red-breasted Merganser (M)
Ruddy Duck (M)
Plain Chachalaca (R)
Mountain Quail (R)
Northern Bobwhite (R)
Scaled Quail (R)
California Quail (R)
Gambel's Quail (R)
Montezuma Quail (R)
Chukar (R)
Himalayan Snowcock (R)
Gray Partridge (R)
Ring-necked
  Pheasant (R)
Ruffed Grouse (R)
Greater Sage-Grouse (R)
Gunnison Sage-Grouse (R)
Dusky Grouse (R)
Sooty Grouse (R)
Greater Prairie-Chicken (R)

Lesser Prairie-Chicken (R)
Wild Turkey (R)
Greater Flamingo (R)
Least Grebe (R)
Pied-billed Grebe (M)
Horned Grebe (M)
Red-necked Grebe (M)
Eared Grebe (M)
Western Grebe (M)
Clark's Grebe (M)
Rock Pigeon (R)
Band-tailed Pigeon (R/M)
White-crowned Pigeon (R)
Red-billed Pigeon (R)
Eurasian Collared-Dove (R)
Spotted Dove (R)
Inca Dove (R)
Common Ground Dove (R)
White-tipped Dove (R)
White-winged Dove (R)
Mourning Dove (R)
Mangrove Cuckoo (R)
Greater Roadrunner (R)
Smooth-billed Ani (R)
Groove-billed Ani (R)
Lesser Nighthawk (M)
Common Poorwill (R/M)
Buff-collared Nightjar (M)
Eastern Whip-poor-will (M)
Common Pauraque (R)
White-throated Swift (M/R)
Rivoli's Hummingbird (M)
Blue-throated
  Mountain-gem (M)

Black-chinned
  Hummingbird (M)
Broad-tailed
  Hummingbird (M)
Broad-billed
  Hummingbird (M)
Buff-bellied
  Hummingbird (R)
Violet-crowned
  Hummingbird (M)
Anna's Hummingbird (R)
Costa's Hummingbird (R/M)
Yellow Rail (M)
Black Rail (M)
Virginia Rail (R/M)
Sora (M)
Ridgway's Rail (R)
Clapper Rail (R/M)
King Rail (M/R)
Purple Gallinule (M/R)
Common Gallinule (M)
American Coot (M/R)
Sandhill Crane (M)
Whooping Crane (M/R)
Limpkin (R)
Black-necked Stilt (M/R)
American Avocet (M)
American Oystercatcher (R)
Black Oystercatcher (R)
Snowy Plover (M/R)
Killdeer (M/R)
Black-bellied Plover (M)
Wilson's Plover (M)
Piping Plover (M)

Mountain Plover (M)
Semipalmated Plover (M)
Long-billed Curlew (M)
Marbled Godwit (M)
Dunlin (M)
Purple Sandpiper (M)
Long-billed Dowitcher (M)
Wilson's Snipe (M)
Wandering Tattler (M)
Lesser Yellowlegs (M)
Willet (M)
Greater Yellowlegs (M)
American Woodcock (M/R)
Great Skua (M)
Pomarine Jaeger (M)
Dovekie (M)
Common Murre (M)
Parakeet Auklet (R/M)
Cassin's Auklet (R)
Horned Puffin (R/M)
Tufted Puffin (M)
Razorbill (R/M)
Marbled Murrelet (R/M)
Scripp's Murrelet (R)
Craveri's Murrelet (R)
Ancient Murrelet (R/M)
Rhinoceros Auklet (M)
Black-legged
  Kittiwake (R/M)
Laughing Gull (R/M)
Heermann's Gull (M/R)
Short-billed Gull (M)
Ring-billed Gull (M)
Western Gull (R)

Yellow-footed Gull (R)
California Gull (M/R)
Herring Gull (M/R)
Iceland Gull (M)
Lesser Black-backed Gull (M)
Glaucous-winged
  Gull (R/M)
Great Black-backed
  Gull (R/M)
Brown Noddy (R)
Sooty Tern (R)
Common Loon (M)
Red-throated Loon (M)
Arctic Loon (M)
Pacific Loon (M)
Bridled Tern (R)
Gull-billed Tern (R/M)
Caspian Tern (M)
Forster's Tern (M/R)
Royal Tern (M/R)
Sandwich Tern (M/R)
Elegant Tern (M)
White-tailed Tropicbird (R)
Red-billed Tropicbird (R)
Leach's Storm-Petrel (R)
Ashy Storm-Petrel (R)
Black Storm-Petrel (R)
Least Storm-Petrel (R)
Northern Fulmar (R/M)
Wood Stork (R)
Magnificent Frigatebird (R)
Northern Gannet (M) ·
Brown Booby (R)
Brandt's Cormorant (R)

APPENDIX

Neotropic Cormorant (R)
Double-crested
  Cormorant (M/R)
Great Cormorant R/M)
Anhinga (R)
American White Pelican (M)
Brown Pelican (R)
American Bittern (M/R)
Least Bittern (M/R)
Great Blue Heron (R/M)
Great Egret (M/R)
Snowy Egret (M/R)
Little Blue Heron (M/R)
Tricolored Heron (R)
Reddish Egret (R)
Western Cattle Egret (M/R)
Green Heron (M/R)
Black-crowned Night
  Heron (M/R)
Yellow-crowned Night
  Heron (M/R)
White Ibis (R)
White-faced Ibis (M/R)
Glossy Ibis (M/R)
Roseate Spoonbill (R)
Turkey Vulture (M/R)
Black Vulture (R)
California Condor (R)
Osprey (M/R)
White-tailed Kite (R)
Hook-billed Kite (R)
Golden Eagle (M/R)
Northern Harrier (M/R)
Sharp-shinned Hawk (M/R)

Cooper's Hawk (R/M)
Bald Eagle (R/M)
Snail Kite (R)
Harris's Hawk (R)
White-tailed Hawk (R)
Red-shouldered
  Hawk (R/M)
Short-tailed Hawk (R)
Red-tailed Hawk (R/M)
Zone-tailed Hawk (M)
Ferruginous Hawk (M/R)
Rough-legged Hawk (M)
Barn Owl (R)
Flammulated Owl (M)
Western Screech-Owl (R)
Eastern Screech-Owl (R)
Whiskered Screech-Owl (R)
Great Horned Owl (R)
Northern Pygmy-Owl (R)
Ferruginous Pygmy-Owl (R)
Elf Owl (M/R)
Burrowing Owl (M/R)
Long-eared Owl (M/R)
Short-eared Owl (M/R)
Spotted Owl (R)
Barred Owl (R)
Northern Saw-whet
  Owl (M/R)
Elegant Trogon (M)
Lewis's Woodpecker (M/R)
Red-headed
  Woodpecker (R/M)
Acorn Woodpecker (R)
Gila Woodpecker (R)

Golden-fronted
  Woodpecker (R)
Red-bellied Woodpecker (R)
Williamson's
  Sapsucker (M/R)
Yellow-bellied Sapsucker (M)
Red-napped
  Sapsucker (M/R)
Red-breasted
  Sapsucker (R/M)
Nuttall's Woodpecker (R)
Ladder-backed
  Woodpecker (R)
Red-cockaded
  Woodpecker (R)
White-headed
  Woodpecker (R)
Arizona Woodpecker (R)
Pileated Woodpecker (R)
Northern Flicker (R/M)
Gilded Flicker (R)
Crested Caracara (R)
American Kestrel (M/R)
Merlin (M/R)
Aplomado Falcon (R)
Prairie Falcon (R)
Peregrine Falcon (M)
Monk Parakeet (R)
Green Parakeet (R)
Red-crowned Parrot (R)
Rose-throated Becard (R)
Northern
  Beardless-Tyrannulet (M/R)
Dusky-capped Flycatcher (M)

Brown-crested Flycatcher (M)
Great Kiskadee (R)
Sulphur-bellied
  Flycatcher (R)
Tropical Kingbird (R)
Couch's Kingbird (R)
Cassin's Kingbird (M/R)
Thick-billed Kingbird (R)
Black-capped Vireo (M)
Blue-gray
  Gnatcatcher (M/R)
Phainopepla (M/R)
Gray Flycatcher (M)
Dusky Flycatcher (M)
Western Flycatcher (M)
Black Phoebe (R)
Eastern Phoebe (M/R)
Say's Phoebe (M/R)
Vermilion Flycatcher (R/M)
Loggerhead Shrike M/R)
Northern Shrike (M/R)
White-eyed Vireo (M/R)
Bell's Vireo (M)
Gray Vireo (M/R)
Hutton's Vireo (R)
Cassin's Vireo (M)
Blue-headed Vireo (M)
Plumbeous Vireo (M)
Brown Jay (R)
Green Jay (R)
Pinyon Jay (R)
Steller's Jay (R)
Blue Jay (R/M)
Florida Scrub-Jay (R)

Island Scrub-Jay (R)
California Scrub-Jay (R)
Woodhouse's Scrub-Jay (R)
Clark's Nutcracker (R)
Black-billed Magpie (R)
Yellow-billed Magpie (R)
American Crow (R/M)
Tamaulipas Crow (R)
Fish Crow (R)
Chihuahuan Raven (R)
Eurasian Skylark (R)
Horned Lark (M/R)
Tree Swallow (M)
Violet-green Swallow (M)
Northern Rough-winged
  Swallow (M/R)
Cave Swallow (M)
Carolina Chickadee (R)
Mountain Chickadee (R)
Mexican Chickadee (R)
Chestnut-backed
  Chickadee (R)
Bridled Titmouse (R)
Oak Titmouse (R)
Juniper Titmouse (R)
Tufted Titmouse (R)
Black-crested Titmouse (R)
Verdin (R)
Bushtit (R)
White-breasted Nuthatch (R)
Pygmy Nuthatch (R)
Brown-headed Nuthatch (R)
Brown Creeper (R/M)
Rock Wren (R/M)

Canyon Wren (R)
House Wren (M/R)
Pacific Wren
Winter Wren (M/R)
Sedge Wren (M)
Marsh Wren (M/R)
Carolina Wren (R)
Bewick's Wren (R/M)
Cactus Wren (R)
California Gnatcatcher (R)
Black-tailed Gnatcatcher (R)
Black-capped
  Gnatcatcher (R)
American Dipper (R)
Red-whiskered Bulbul (R)
Red-vented Bulbul (R)
Golden-crowned
  Kinglet (M/R)
Ruby-crowned
  Kinglet (M/R)
Wrentit (R)
Eastern Bluebird (M/R)
Western Bluebird (M/R)
Mountain Bluebird (M/R)
Townsend's Solitaire (M/R)
Hermit Thrush (M/R)
American Robin (M/R)
Clay-colored Thrush (R)
Varied Thrush (M/R)
Gray Catbird (M/R)
Curve-billed Thrasher (R)
Brown Thrasher (M/R)
Long-billed Thrasher (R)
Bendire's Thrasher (M/R)

California Thrasher (R)

LeConte's Thrasher (R)

Crissal Thrasher (R)

Sage Thrasher (M)

Northern Mockingbird (R)

European Starling (R)

Common Myna (R)

Crested Myna (R)

Common Hill Myna (R)

Cedar Waxwing (R/M)

Olive Warbler (R)

House Sparrow (R)

Eurasian Tree Sparrow (R)

American Pipit (M)

Sprague's Pipit (M)

Gray-crowned
  Rosy-finch (M/R)

Black Rosy-finch (R)

Brown-capped Rosy-finch (R)

House Finch (R)

Purple Finch (M/R)

Cassin's Finch (R/M)

Evening Grosbeak

Cassia Crossbill (R)

Pine Siskin (M/R)

Lesser Goldfinch (M/R)

Lawrence's Goldfinch (M/R)

American Goldfinch (R/M)

Lapland Longspur (M)

Chestnut-collared
  Longspur (M)

Smith's Longspur (M)

Thick-billed Longspur (M)

Olive Sparrow (R)

Green-tailed Towhee (M/R)

Spotted Towhee (R/M)

Eastern Towhee (M/R)

Rufous-crowned Sparrow (R)

Canyon Towhee (R)

California Towhee (R)

Abert's Towhee (R)

Rufous-winged Sparrow (R)

Cassin's Sparrow (R/M)

Botteri's Sparrow (R)

Bachman's Sparrow (R/M)

American Tree Sparrow (M)

Chipping Sparrow (M/R)

Clay-colored Sparrow (M)

Brewer's Sparrow (M/R)

Field Sparrow (R/M)

Black-chinned
  Sparrow (M/R)

Vesper Sparrow (M/R)

Lark Sparrow (M/R)

Five-striped Sparrow (R)

Black-throated
  Sparrow (M/R)

Sagebrush Sparrow (M/R)

Bell's Sparrow (R)

Lark Bunting (M/R)

Savannah Sparrow (M/R)

Grasshopper Sparrow (M/R)

Baird's Sparrow (M)

Henslow's Sparrow (M)

LeConte's Sparrow(M)

Seaside Sparrow (R/M)

Nelson's Sparrow (M)

Saltmarsh Sparrow (M/R)

Fox Sparrow (M/R)

Song Sparrow (R/M)

Lincoln's Sparrow (M)

Swamp Sparrow (M/R)

White-throated
  Sparrow (M/R)

White-crowned
  Sparrow (M/R)

Golden-crowned
  Sparrow (M)

Harris's Sparrow (M)

Dark-eyed Junco (M/R)

Yellow-eyed Junco (R)

Spot-breasted Oriole (R)

Altamira Oriole (R)

Audubon's Oriole (R)

Yellow-headed
  Blackbird (M/R)

Eastern Meadowlark (R/M)

Western Meadowlark (R/M)

Chihuahuan Meadowlark (R)

Red-winged
  Blackbird (M/R)

Tricolored Blackbird (R)

Shiny Cowbird (R)

Bronzed Cowbird (R)

Brown-headed
  Cowbird (M/R)

Rusty Blackbird (M)

Brewer's Blackbird (M/R)

Common Grackle (M/R)

Boat-tailed Grackle (R)

Great-tailed Grackle (R)

Black-and-white Warbler (M)

Prothonotary Warbler (M)

Cape May Warbler (M)

Northern Parula (M)

Tropical Parula (R)

Palm Warbler (M)

Pine Warbler (M/R)

Yellow-rumped
  Warbler (M/R)

Yellow-throated
  Warbler (M/R)

Prairie Warbler (M/R)

Grace's Warbler (M)

Black-throated Gray
  Warbler (M)

Townsend's Warbler (M/R)

Black-throated Green
  Warbler (M)

Wilson's Warbler (M)

Painted Redstart (M)

Orange-crowned
  Warbler (M/R)

Colima Warbler (R)

Lucy's Warbler (M)

Nashville Warbler (M)

Common
  Yellowthroat (M/R)

Kirtland's Warbler (M)

Hepatic Tanager M)

Northern Cardinal (R)

Pyrrhuloxia (R)

Lazuli Bunting (M)

Varied Bunting (M/R)

Painted Bunting (M)

Morelet's Seedeater (R)

Tier 3: Intermediate Migrants

Migrating birds wintering as far north as southern Mexico, the Greater Antilles, Central America, and extreme northern South America. While the Greater Antilles are only 100 miles south of Florida; species wintering there are, here, classified as intermediate migrants because the Greater Antilles lie at approximately the same latitude as the Yucatan Peninsula (a region designated here as part of Central America).

Least Tern	Least Flycatcher
Common Tern	Hammond's Flycatcher
Chuck-will's-widow	Yellow-throated Vireo
Mexican Whip-poor-will	Philadelphia Vireo
Vaux's Swift	Black-whiskered Vireo
Ruby-throated Hummingbird	Warbling Vireo
Rufous Hummingbird	Bank Swallow
Allen's Hummingbird	Barn Swallow
Calliope Hummingbird	Bicknell's Thrush
Lucifer Hummingbird	Swainson's Thrush
Black Turnstone	Wood Thrush
Least Sandpiper	Yellow-breasted Chat
Western Sandpiper	Baltimore Oriole
Short-billed Dowitcher	Hooded Oriole
Sabine's Gull	Orchard Oriole
Bonaparte's Gull	Bullock's Oriole
Gray Hawk	Scott's Oriole
Common Black Hawk	Ovenbird
Ash-throated Flycatcher	Worm-eating Warbler
Great-crested Flycatcher	Louisiana Waterthrush
Western Kingbird	Northern Waterthrush
Scissor-tailed Flycatcher	Blue-winged Warbler
Greater Pewee	Yellow Warbler
Yellow-bellied Flycatcher	Swainson's Warbler
Willow Flycatcher	Tennessee Warbler

Magnolia Warbler
Chestnut-sided Warbler
Black-throated Blue Warbler
Hermit Warbler
Golden-cheeked Warbler
Virginia's Warbler
Red-faced Warbler
MacGillivray's Warbler
Mourning Warbler
Kentucky Warbler

Hooded Warbler
American Redstart
Western Tanager
Summer Tanager
Rose-breasted Grosbeak
Black-headed Grosbeak
Blue Grosbeak
Indigo Bunting
Dickcissel

Tier 4: Long Distance Migrants

Birds wintering in middle and southern South America, the South Pacific, Africa, Southeast Asia, and southern-hemisphere ocean waters.

Broad-winged Hawk
Yellow-billed Cuckoo
Black-billed Cuckoo
Common Nighthawk
Antillean Nighthawk*
Black Swift
Chimney Swift
American Golden-Plover
Pacific Golden-Plover
Upland Sandpiper
Bristle-thighed Curlew
Whimbrel
Bar-tailed Godwit
Hudsonian Godwit

Ruddy Turnstone
Red Knot
Surfbird
Stilt Sandpiper
Sanderling
Baird's Sandpiper
White-rumped Sandpiper
Buff-breasted Sandpiper
Pectoral Sandpiper
Semipalmated Sandpiper
Spotted Sandpiper
Solitary Sandpiper
Wilson's Phalarope
Red-necked Phalarope

* Migration routes and wintering destination remain unknown

Red Phalarope
Parasitic Jaeger
Long-tailed Jaeger
Franklin's Gull
Aleutian Tern
Black Tern
Roseate Tern
Arctic Tern
Fork-tailed Storm-Petrel
Swallow-tailed Kite
Mississippi Kite
Swainson's Hawk
Eastern Kingbird
Gray Kingbird
Olive-sided Flycatcher
Western Wood-Pewee
Eastern Wood-Pewee
Acadian Flycatcher
Alder Flycatcher
Red-eyed Vireo

Purple Martin
Cliff Swallow
Arctic Warbler
Bluethroat
Northern Wheatear
Veery
Gray-cheeked Thrush
Eastern Yellow Wagtail
White Wagtail
Red-throated Pipit
Bobolink
Golden-winged Warbler
Cerulean Warbler
Bay-breasted Warbler
Blackburnian Warbler
Blackpoll Warbler
Canada Warbler
Connecticut Warbler
Scarlet Tanager

BIBLIOGRAPHY

American Ornithologists' Union. *Checklist of North American Birds*, 7th ed. Washington, DC: American Ornithologists' Union, 1998.

Angulate, J. L., and R. Galati. "Golden-crowned Kinglet (*Regulus satrapa*)." In *The Birds of North America*. A. Poole and F. Gill, eds. Philadelphia, PA: The Academy of Natural Sciences, 1997.

Bechard, M. J., and T. R. Swem. "Rough-legged Hawk (*Buteo lagopus*)." In *The Birds of North America*. 2002.

"Black-capped Chickadee." In *The Birds of North America*.

Choate, Ernest A. "The Cold Winter—1977." *Cape May Geographic*, November 8, 2013. https://capemaygeographic.wordpress.com/2013/11/08/the-cold-winter.

Conway, C. J. "Virginia Rail (*Rallus limicola*)." In *The Birds of North America*. 1995.

Cornell Lab of Ornithology. *All About Birds*. http://www.allaboutbirds.org.

Del Hoyo, Josep, et al. *Handbook of the Birds of the World*, vol. 1–16. Barcelona: Lynx Editions, 2011.

Ficken, M. S., M. A. McLaren, and J. P. Hallman. "Boreal Chickadee (*Parus hudsonicus*)." In *The Birds of North America*. 1996.

Gill, Frank B. *Ornithology* 2nd ed. New York: W. H. Freeman and Co., 1995.

Hatch, J. J. "Arctic Tern (*Sterna paradisea*)." In *The Birds of North America*. 2002.

Holder, K., and R. Montgomerie. "Rock Ptarmigan (*Lagopus mucus*)." In *The Birds of North America*. 1993.

Jones, P. W., and T. M. Donovan. "Hermit Thrush (*Catharus guttatus*)." In *The Birds of North America*. 1996.

Kren, J., and A. C. Zoerb. "Northern Wheatear (*Oenanthe oenanthe*)." In *The Birds of North America*. 1997.

Lyon, B., and R. Montgomerie. "Snow Bunting and McKay's Bunting (*Plectrophenax nivalis* and *P. Hyperboreus*)." In *The Birds of North America*. 1995.

Madge, Steve, and Phil McGowan. *Pheasants, Partridges, & Grouse: Including Buttonquails, Sandgrouse, and Allies*. Princeton, NJ: Princeton University Press, 2002.

Payne, L. X., and E.P. Pierce. "Purple Sandpiper (*Calidris maritima*)." In *The Birds of North America*. 2002.

Peterson, Roger T. *A Field Guide to the Birds: Giving Field Marks of All Species Found in Eastern North America*. Cambridge, MA: The Riverside Press. 1934.

Richardson, Rachel M., et al. "Rapid Population Decline in McKay's Bunting, an Alaskan Endemic, Highlights the Species' Current Status Relative to International Standards for Vulnerable Species." *Ornithological Applications* 126, no. 2 (May 2024): https://doi.org/10.1093/ornithapp/duad064.

Robbins, Jim. "The Wonder Bird." *Smithsonian*, January/February 2022. https://www.smithsonianmag.com/science-nature/hudsonian-godwit-flies-thousands-miles-without-resting-180979248.

Robertson, G. J., and R. I. Goudie. "Harlequin Duck (*Histrionicus histrioicus*)." In *The Birds of North America*. 1999.

Salabanks, R., and F.C. James. "American Robin (*Turdus migratorius*)." In *The Birds of North America*. 1999.

Sibley, David A. *The Sibley Guide to Birds*. 2nd ed. New York: Alfred A. Knopf, 2014.

ABOUT THE AUTHORS

Kevin Karlson

PETE DUNNE is an author and the founder of the World Series of Birding, former director of natural history information for the New Jersey Audubon Society, and former director of the Cape May Bird Observatory. An experienced tour leader, he is well known for his columns and contributions to publications such as *American Birds* and *Birding*. He is also the author of more than twenty books, including *Birds of Prey*, *Gulls Simplified*, *The Art of Pishing*, *The Wind Masters*, and, most recently, *The Shorebirds of North America*.

DAVID ALLEN SIBLEY is the author and illustrator of several successful nature guides, including *The Sibley Guide to Birds*. He has contributed to *Smithsonian*, *Science*, *The Wilson Journal of Ornithology*, *Birding*, *BirdWatching*, *North American Birds*, and the *New York Times*. He is the recipient of the Roger Tory Peterson Award for lifetime achievement from the American Birding Association and the Linnaean Society of New York's Eisenmann Medal.

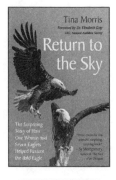